T0220477

Digital Data Collection and Information Privacy Law

In *Digital Data Collection and Information Privacy Law*, Mark Burdon argues for the reformulation of information privacy law to regulate new power consequences of ubiquitous data collection. Examining developing business models based on collections of sensor data – with a focus on the 'smart home' – Burdon demonstrates the challenges that are arising for information privacy's control model and its application of principled protections of personal information exchange. By reformulating information privacy's primary role of individual control as an interrupter of modulated power, Burdon provides a foundation for future law reform and calls for stronger information privacy law protections. This book should be read by anyone interested in the role of privacy in a world of ubiquitous and pervasive data collection.

Mark Burdon is Associate Professor of Law at Queensland University of Technology. His research interests include the regulation of information security practices, legislative frameworks for mandatory reporting of data breaches and the onset of a 'sensor society'. Mark's most recent works focus on privacy issues arising from smart homes, particularly those involving domestic violence reporting and smart home insurance.

Cambridge Intellectual Property and Information Law

As its economic potential has rapidly expanded, intellectual property has become a subject of front-rank legal importance. Cambridge Intellectual Property and Information Law is a series of monograph studies of major current issues in intellectual property. Each volume contains a mix of international, European, comparative and national law, making this a highly significant series for practitioners, judges and academic researchers in many countries.

Series Editors

Lionel Bently
Herchel Smith Professor of Intellectual Property Law, University of Cambridge

Graeme Dinwoodie
Global Professor of Intellectual Property Law, Chicago-Kent College of Law, Illinois Institute of Technology

Advisory Editors

William R. Cornish, *Emeritus Herchel Smith Professor of Intellectual Property Law, University of Cambridge*

François Dessemontet, *Professor of Law, University of Lausanne*

Jane C. Ginsburg, *Morton L. Janklow Professor of Literary and Artistic Property Law, Columbia Law School*

Paul Goldstein, *Professor of Law, Stanford University*

The Rt Hon. Sir Robin Jacob, *Hugh Laddie Professor of Intellectual Property, University College London*

Ansgar Ohly, *Professor of Intellectual Property Law, Ludwig-Maximilian University of Munich*

A list of books in the series can be found at the end of this volume.

Digital Data Collection and Information Privacy Law

Mark Burdon

Queensland University of Technology

CAMBRIDGE
UNIVERSITY PRESS

CAMBRIDGE
UNIVERSITY PRESS

University Printing House, Cambridge CB2 8BS, United Kingdom

One Liberty Plaza, 20th Floor, New York, NY 10006, USA

477 Williamstown Road, Port Melbourne, VIC 3207, Australia

314-321, 3rd Floor, Plot 3, Splendor Forum, Jasola District Centre, New Delhi - 110025, India

103 Penang Road, #05-06/07, Visioncrest Commercial, Singapore 238467

Cambridge University Press is part of the University of Cambridge.

It furthers the University's mission by disseminating knowledge in the pursuit of education, learning and research at the highest international levels of excellence.

www.cambridge.org
Information on this title: www.cambridge.org/9781108406017
DOI: 10.1017/9781108283717

First published 2020
First paperback edition 2022

A catalogue record for this publication is available from the British Library

Library of Congress Cataloging in Publication data
Names: Burdon, Mark, 1967- author.
Title: Digital data collection and information privacy law / Mark Burdon,
Queensland University of Technology.
Description: Cambridge, United Kingdom ; New York, NY, USA : Cambridge
University Press, 2020. | Series: Cambridge intellectual property and
information law | Includes bibliographical references and index.
Identifiers: LCCN 2019036041 | ISBN 9781108417921 (hardback) | ISBN
9781108283717 (ebook)
Subjects: LCSH: Data protection--Law and legislation. | Databases--Law and
legislation. | Big data--Social aspects. | Technological
innovations--Social aspects. | Internet of things--Social aspects.
Classification: LCC K3264.C65 B87 2020 | DDC 342.08/58--dc23
LC record available at https://lccn.loc.gov/2019036041

ISBN 978-1-108-41792-1 Hardback
ISBN 978-1-108-40601-7 Paperback

This book is dedicated to Tom,
Sophie and Sally. Love always.

Contents

Figure and Tables

Figure

Tables

Acknowledgements

As with any large piece of work, this book has been a few years in the making – a bit longer than expected – and I thank Matt Gallaway and Cambridge University Press for being patient and flexible. I also very much thank Laura Blake at CUP and Priyaa Menon at Lumina for their patience and understanding.

I hadn't quite appreciated how many years I had been working on the book until it came to the final drafting spree and I returned to a topic that has fascinated me for the best part of a decade now – the relationship between privacy and power. Returning to previous ideas, with a fresh perspective, also gave me the opportunity to reflect on the book's journey. During that time, I have changed institutions and returned to the Queensland University of Technology (QUT), having spent eight years at The University of Queensland (UQ).

I would like to thank the deans and heads of school of both institutions who have supported me in this journey: Ross Grantham, for always leaving his door open for intellectual ponderings and being very generous in reading the early genesis of the book; the Hon. Sarah Derrington, president of the Australian Law Reform Commission, for believing in me and supporting me during her time as dean of UQ Law; John Humphrey, for providing me the opportunity to return to QUT Law; and, last but not least, Judith McNamara, for giving me the time and space to finish the book.

I have also been fortunate to work with a wonderful array of different academics at both institutions. I've gained so much from my interactions with them that they likely don't realise the impact of their thoughtful considerations over coffee and during corridor chats. At UQ, these academics include Russell Hinchy, Heather Douglas, Brad Sherman, Kit Barker, Allison Fish, Rob Mullins, Nick Aroney, Anthony Cassimatis, Barbora Jedlickova, Justine Bell-James, Paul Harpur, Peter Billings, Craig Forrest and many others, particularly the Level 4 gang. I would also like to thank Simon Bronitt for conversations about future research directions that I'm only now starting to fully appreciate.

Finally, I am particularly grateful to Andreas Schloenhardt for research mentoring and friendship. At QUT, Nic Suzor, Kieran Tranter, Bridget Lewis, Richard Johnstone and Afshin Akhtar-Khavari have been fantastic colleagues, both old and new. I am particularly grateful to Nic, who undertook the lion's share of joint teaching responsibilities to give me space to finish the book. You are a gentleman, my friend.

A special note also to my PhD students Brydon Wang and Amy Dunphy, who travelled with me from UQ to QUT: thank you for your trust.

I would also like to thank Simon Lamb, Claudia Levings and Tom Mackie for excellent research assistant support. I have had the pleasure of working with Tom for the best part of two years, and I wish him well in his future journey.

Some of the research for this book was conducted as part of a consultancy with Suncorp's then Strategy and Innovation (SI) team during my time at UQ. I am grateful to Kirsten Dunlop for giving me the opportunity to work with SI. I am especially grateful to Faith Kimani for the many thoughtful chats, including numerous Post-it notes and whiteboard scribblings. Special thanks as well go to Hannah Driscoll for being the consummate organiser. I also had the opportunity to work with many other folks from the SI team, and I am grateful for their collective and individual wisdom. Chapter 10's coverage of institutional tinkering is dedicated to you all, as the tinkerers extraordinaire.

Chapters 5 and 6 are partly derived from short extracts from some of my journal articles: Mark Burdon and Paul Harpur, 'Re-conceptualising Privacy and Discrimination in an Age of Talent Analytics' (2014) 37 University of New South Wales Law Journal 679; Mark Burdon and Alissa McKillop, 'The Google Street View Wi-Fi Scandal and Its Repercussions for Privacy Regulation' (2013) 39 Monash University Law Review 702; Mark Burdon, 'Contextualizing the Tensions and Weaknesses of Data Breach Notification and Information Privacy Law' (2010) 27 Santa Clara Computer and High Technology Law Journal 63; Mark Burdon, 'Privacy Invasive Geo-mashups: Privacy 2.0 and the Limits of First Generation Information Privacy Laws' [2010] University of Illinois Journal of Law, Technology and Policy 1.

Finally, I thank Sally, Tom and Sophie for their patience, love and support. Writing can be an isolating activity, but never a lonely one with the three of you in my life. Love you always.

1   Introduction

Let's start our journey with a question – a simple one: what have you done today? Take a second to think about it before you read on.

You have probably got out of bed as a starter. How did you sleep? You have probably already prepared for your day. Maybe you are the early-starter type, so you are straight into the shower, dressed and off to the kitchen to make breakfast. What did you have for breakfast? Perhaps something that involved fishing in a fridge or using a kitchen device such as a kettle, a coffee machine or a toaster. You may have younger mouths to feed also, nappies to change or school lunch boxes to prepare – all before you head off to work, wherever and in whatever form that may take. And as you progress during your day, you have the bings and beeps of your smartphone as you negotiate your life and try to understand the world around you. Maybe you need to order a taxi or an Uber. Maybe you need to shop or withdraw cash. Maybe you need to make a call or send a text, attend a meeting or pick up the kids – or the many other things that now pack our expected lives.

Now let me ask you this: what did you do this day last week? Possibly similar activities but maybe something different. What about this date last month or even last year? It gets harder to distinguish one day from the next, unless something of note has occurred.

But what if we lived in a world where data about everything was collected? What if all devices, all buildings and all infrastructures were sensorised?

It could be a smarter world. We would know more about our own patterns and foibles, as we could easily compile a historical record of all our activities. We would no longer struggle to work out what we did today, last week, last month or anytime previously. It would all be recorded: the times we boil the kettle for our morning cup of tea, how we travel to work and when we travel – maybe even the days we feel sad, indicated by the movies we watch or music we listen to.

Smarter still, these types of sensor data can also be used collectively and not just individually. So, your repetitive kettle use, at the same time each morning, can be fed into electricity usage across the broader population's grid. We can work out travel patterns of communities or populations by tracking individual smartphones to better allocate resources. We can even consider the mental health of individuals in seemingly massive populations by monitoring social media and activities in other public fora.

'Smartness' in the smart world thus connotes prediction and prescription: prediction, in that we can better assess how individuals, communities and populations will act, behave and possibly even feel; prescription, in that we can target outcomes to better meet predicted futures.

The smart world obliquely consists of pervasive collections of data from a vast range of sensorised devices, infrastructures and environments. It puts us on a path where data about everything could be collected for smartness to function and flourish. The success of the smart world is consequently dependent upon vast scales of data collection, particularly from sensorised devices. The smart world is therefore the 'collected' world. But what does it mean for us to live in a collected world? How do we make sense of it?

1.1   The Book's Thesis

A common frame for addressing these questions is information privacy or data protection law because it directly governs our relationship with the many data collectors of the smart and collected worlds. Different concepts of information privacy exist, but its traditional and dominant form regards individual control over personal information. We have an intrinsic say in what personal information is collected about us, how that information is subsequently used and how we can access or correct that information from data-collecting organisations.

Information privacy law provides a range of life-cycle protections that begin at the point of data collection and end with destruction or de-identification of no-longer-required data. In the interim, data collection organisations have a range of obligations to fulfil: the individual should be notified about the purposes of collection so that they can meaningfully consent to subsequent uses. Personal information can generally only be used for a defined purpose about which the individual is adequately informed. Individuals have a range of interaction mechanisms that seek to ensure the maintenance of control by giving them the ability to affirm the accuracy and currency of collected personal information. Personal information, once collected and stored, must be kept secure.

Despite the commonality of these types of protections, which originate from the same founding roots of principled application, different jurisdictions have different emphases about how information privacy law should apply. Two foundational issues are of importance: (1) the type of information that will trigger legislative or regulatory response and (2) the existence or absence of principled protections at the point of data collection. These issues are at the crux of significant variations between the jurisdictional approaches of the European Union (EU), the United States and the non-EU Organisation for Economic Co-operation and Development (OECD) countries such as Australia.

The ubiquitous collections of sensor data now unfolding in the collected world fundamentally challenge traditional information privacy protections. It is not legally clear whether, and in what forms, sensor data will be classified as the types of information that trigger information privacy law's application. Sensor data that can be collected from many sources can now be used to infer patterns and behaviours, thus obviating the need to collect personal or sensitive information directly from us. Moreover, the pervasive and continuous harvesting of all data trails challenges our conscious ability to rationally consent to the process of personal information collection.

The new norm of sensorised data collection processes, at the heart of developing smart business models, is also very different to the past collections of personal information that information privacy law is based upon. In the new forms of collections, we move from being active and autonomous participants who can control our own data to passive supplicants of infrastructures that are so unfathomable that the idea of individual control is nothing but an illusion. Instead of shaping our own destinies, we could instead be shaped by the many and multifarious organisations that collect, analyse and reuse our data.

To what extent does the collected world therefore require us to rethink the conceptual basis of control-based information privacy and its manifestation as the principled protections of information privacy law?

This book seeks to address that question. It argues that the collected world's new norms of sensor data collection are such that we need to reconsider the control basis of information privacy and its application in law. Several reasons are addressed.

The one-to-one, controllable data collection exchanges between individuals and data collectors are increasingly becoming a thing of the past. Collections from the sensorised spaces of the collected world, such as the smart home, involve multiple data collectors that collect sensor data from different sources and pathways. A new power context emerges relating to the ability to coalesce data from smart environments, which are so fragmented that it is difficult for any one party to have control. Like the notions of the smart and collected worlds, the abilities of data collection and sense-making are intrinsically linked. This is recognised in information privacy law regarding limitations on primary uses and resultant secondary purposes. However, these limitations focus on the ability of an individual to retain control over their information as a right of input into the decision-making capacities of data-collecting organisations. Thus, information privacy law seeks to ameliorate power imbalances, but it does so in a way that no longer matches the accumulation and use of power in the sensorised environments of the collected world.

Fundamentally linked to the new processes of sensorised data collections are the underlying business models that contribute to the relentless push for truly ubiquitous collections. The drive is most clearly articulated in descriptions of big data analytics and the desire to collect everything to find insight through the unintuitive. While the force of the driver may have diminished somewhat, the after-effects are still palpable due to the norm of sensorised collections that continue to pervade. Sensor data can provide insight into historical patterns of activities, behaviours and personalities that can be used to predict and prescript future service outcomes. Value resides in the ability to locate, collect and analyse sensor data, which has now become the desired target for the new business models of the collected world.

In considering the new forms of sensor collection and business models, it becomes increasingly important to reconsider what control and autonomy means in the context of information privacy protections. It is arguable that control of personal information is no longer attainable in the sensorised environments of the collected world that have so many points of interaction across multiple parties. Even if it is attainable, it is questionable whether being able to control personal information provides the type of protection that is required. Individual control is more likely to operate effectively when there is a limit to the data that is generated and collected. However, it is questionable whether individual control is an effective measure in the face of ubiquitous collections of sensor data from complex, fractured and contested environments that are increasingly boundary-confused.

As a result, this book argues that both the concept of information privacy and its application in law need to be reconsidered because of the nature of the collected world. Section 1.2 reveals how this reconsideration unfolds throughout the book's three parts and nine substantive chapters.

1.2   The Book's Coverage

The book argues for a reformulation of information privacy's control basis to one that explicitly regards the new power consequences of the collected world. In so doing, Part I outlines the scope of the collected world. It examines developing business models, based on collections of sensor data, as a means of demonstrating the challenges that will arise for information privacy's control model and the application of principled protections of personal information exchange. Part II outlines the conceptual basis of information privacy and its expression as information privacy law. Part III then details the challenges to both the concept and the application of information privacy law that arise from the collected world, using the smart home coverage of Part I and the information privacy coverage of Part II. Next, in Part III, Julie Cohen's work is used in the reformulation of information privacy's new role of interrupting modulated power.[1] Part III's reformulation also provides a foundation for some areas of future information privacy law reform, specifically focusing on the type of information that triggers regulatory intervention and the need for a strong collection principle based on fairness. Throughout, the book's focus is on private sector data collectors rather than public, given the commercial drive for the collected world.

1.2.1   Part I: The Collected World

Part I examines the scope of the collected world, and it contends that the smart world is a collected world. The frame of the collected world is, by its very nature, broad. An understanding of conceptual breadth is necessary to fully consider the power consequences that unfold in

---

[1] Julie E Cohen, *Configuring the Networked Self: Law, Code, and the Play of Everyday Practice* (Yale University Press 2012); Julie E Cohen, 'What Privacy Is For' (2013) 126 Harvard Law Review 1904; Julie E Cohen, 'Between Truth and Power' in M Hildebrandt and Bibi van den Berg (eds), *Freedom and Property of Information: The Philosophy of Law Meets the Philosophy of Technology* (Routledge 2014); Julie E Cohen, 'The Networked Self in the Modulated Society' in Wouter de Been, Payal Arora and Mireille Hildebrandt (eds), *Crossroads in New Media, Identity and Law: The Shape of Diversity to Come* (Palgrave Macmillan 2015); Julie E Cohen, 'The Surveillance-Innovation Complex: The Irony of the Participatory Turn' in Darin Barney and others (eds), *The Participatory Condition in the Digital Age* (Minnesota University Press 2016); Julie E. Cohen, 'Affording Fundamental Rights: A Provocation Inspired by Mireille Hildebrandt' (2016) 4 Critical Analysis of Law 78; Julie E Cohen, 'The Biopolitical Public Domain: The Legal Construction of the Surveillance Economy' (2018) 31 Philosophy & Technology 213; Julie E Cohen, 'Turning Privacy Inside Out' (2019) 20 Theoretical Inquiries in Law 1; Julie E. Cohen, 'Review of Zuboff, Shoshana. 2019. The Age of Surveillance Capitalism: The Fight for a Human Future at the New Frontier of Power' (2019) 17 Surveillance & Society 240.

the final chapters of the book, based on Cohen's modulation and her biopolitical domain. However, it is also necessary to consider the finer points of sensor data collection to clearly articulate the collected challenges that will arise for information privacy law. Thus, Part I uses the smart home as a collected-world case study to examine technological infrastructures, the effect of sensorisation and the business models that flow from the new data generation pathways of the collected world.

Chapter 2 begins the scoping discussion and provides an overview of technological development covering three areas – namely, smart individuals, smart buildings and smart environments. A key element of smartness across all these areas is sensorisation and the rapid spread of sensorised devices, which enables new forms of data collection. Focus is placed on the smartphone in its use as a sensor-collecting device that has individual, communal and societal purposes. Chapter 2 concludes by demonstrating the intractable link between the notions and benefits of smartness and what this actually entails in relation to data collection.

Chapter 3 examines the implications of sensorisation in a discrete, collected-world environment – the smart home. This venue is chosen as a collected-world case study for several reasons. First, the smart home is a site of dense sensorisation and thus a good space in which to explore technological infrastructures that underpin the smart world. Second, it is one of the prime sites of sensor data commercialisation, including by the new business models that are developing. Third, the home is a legally protected idyll of the 'private', and it plays a cherished role as a space of autonomous individual growth in liberal societies. The smart home therefore provides an appropriate site in which to consider the consequences of sensorisation, the business models that are unfolding and the impact these models have on both the concept and the application of information privacy law.

Chapter 3 details the complex data generation anatomy of the smart home and examines it from the standpoint of its sensing, reasoning and intervening processes. It does so from a multidisciplinary perspective involving legal, technical and sociological considerations. Even though the smart home is framed as a space of seamless technological experience, its infrastructural anatomy is fragmented and multifarious because it includes multiple data collection devices and diverse collection pathways. Chapter 3 concludes by highlighting that sensor data is key to the operation of the smart home and the business models that are now starting to develop.

Chapter 4 then examines in greater depth the commercial imperatives for collections of sensor data by exploring the rapid development of smart home insurance business models. A brief history of smart home insurance is provided to situate three conceptualised models of smart home data exchange partnership involving insurers and smart home device or system providers – entitled Partnered Data Acquisition, Partnered Intermediary and Platform Entity models.

Each model involves a partnership arrangement between an established insurer and a smart home device or system provider. However, while each model seeks to capitalise value from smart home sensor data, the models do so in different ways, across three spectra – namely, *collection* of data, *connection* to mutually beneficial services and *condition setting*. An analysis of relevant privacy policies is undertaken to highlight each model's operational data structure, the sensor data collected and its foundational characteristics. In turn, this analysis highlights the commercial uses of smart home sensor data involving new logics, business relationships and intended service outcomes. Attention is given to the Platform Entity model specifically, as it harks to many of the power-related issues covered in Part III.

1.2.2   Part II: Information Privacy Law's Concepts and Application

Chapter 4's analysis of relevant privacy policies highlights some key differences in jurisdictional approaches to information privacy law and begins to give insight into some of the challenges to be faced in a collected world. Part II, in preparation for a critical examination of these challenges in Part III, details the conceptual bases of information privacy and its implementation in information privacy law. Part II examines what information privacy law seeks to protect and how it provides protections. On its face, this sounds like a straightforward task, but it is complex due to the many different purposes and emphases required of privacy and information privacy law.

It should already be growing apparent that the book adopts certain language and positioning about information privacy law and its relations to broader legal constructs of privacy. In this volume, 'information privacy' and 'data protection laws' are treated synonymously. However, the moniker 'information privacy' is preferred. The preference partly reflects jurisdictional bias, as it is the preferred referential mode in Australia, but it is also preferred because of its explicit acknowledgement of broader aspects of privacy. Information privacy is thus considered a constituent part of privacy law's wider conceptual family. In that regard, the book does not adopt a clean differentiation between information privacy law and privacy law in general, unlike the clearer delineation that is sometimes made between data protection and other areas of privacy law.[2] Again, this is partly down to jurisdictional bias, but it also regards the conceptual examination of the book, which covers information privacy's role as a protector of purportedly autonomous spaces. Once in those conceptual spaces, it becomes more difficult to treat information privacy separately from its broader family.

With that in mind, Chapter 5 identifies, through coverage of key authors, four conceptual themes that underpin information privacy's development: (1) individual control over personal information, (2) informational access and personal autonomous growth, (3) a broader social and relational context and (4) privacy as a structural problem of power. Chapter 5 highlights that the control concept is dominant, as evident by its implementation as information privacy law. However, all themes become important to Part III's analysis, which uses Cohen's work to critique both the control and the autonomy themes as a basis for arguing for a stronger role for information privacy's relational context and power-related elements.

Chapter 6 identifies the different foundational structures and jurisdictional perspectives of information privacy law that involve EU, US and Australian legal frameworks. A historical perspective of information privacy law developments in each jurisdiction is provided based on three founding legal instruments for each jurisdiction. Historical development is important because it highlights that although jurisdictional laws are based on the same principled approach, different jurisdictions adopt different emphases. Two particular emphases are examined: (1) the type of regulated information that triggers regulatory response – namely, personally identifiable information in the United States, personal data in the EU and personal information in Australia – and (2) information privacy law's principled process of protection. Attention is given to collection principles as a means of outlining foundational differences between sectoral and comprehensive regimes of information privacy, particularly regarding the overt use of a notice-and-consent model.

1.2.3   Part III: Information Privacy Law for a Collected Future

Let's recap. Part II sets the conceptual and legal framework from which the collected-world consequences of Part I's analysis can be examined, particularly in relation to sensor data collections from the smart home. Part III then examines the consequences of the collected world and the challenges it will bring for information privacy law's dominant control

---

[2] See Gloria González Fuster, *The Emergence of Personal Data Protection as a Fundamental Right of the EU* (Springer 2014) for an admirable example.

model and its manifestation in information privacy laws predicated on process protections. In doing so, Part III puts forward a reformulated role for information privacy law based on Cohen's work of the past decade, most notably on modulated power.

Chapter 7 returns to the smart home as a way of examining the collected-world challenges that will arise for information privacy law. It highlights that sensor data collections are different in nature to the type of data collections envisaged by the control model of information privacy. Smart home sensor data collections are circular and continuous, which challenges the basis of rationality modes of consent provision. Moreover, the very notion of control is challenged in boundary-dispersed environments, such as the smart home, which are essentially fragmented and contested. These factors give rise to significant challenges for the control model of information privacy law and its focus on the process of personal information exchange.

Chapter 8 then returns to the conceptual analysis of Chapter 5 to address what information privacy *should* protect in a collected world. A sustained critique of the control model is put forward in relation to five intended outcomes of information privacy law. Out of these five intended outcomes, two are broad in nature and reflect the conceptual intentions of information privacy. The first is the enhancement of individual autonomy through the creation of non-interference protections at the point of data collection to ensure unfettered decision making. The second regards the amelioration of power imbalances using power vacuums that preserve spaces for autonomous decision making.

Three narrower outcomes, focused on the application of information privacy law, then supplement the broader conceptual intentions. The first outcome reflects information privacy law's mode of transactional operation, which generally considers provisions of personal information as tradable exercises. The second outcome then regards the use of privacy policies, as information disclosure mechanisms, to ameliorate information asymmetries in transactional modes of operation. This outcome regards information privacy law's inbuilt balancing mechanism, which seeks to secure fair outcomes for individuals while ensuring that data exchanges flow for the benefit of data collection organisations and society.

Chapter 8 introduces Cohen's work as a means of further critiquing the control model and reshaping a conceptual focus of information privacy based on a more explicit, power-related role. The new focus sees a shift in what information privacy seeks to do and in doing so challenges some of the fundamental precepts of the control model and what information privacy currently seeks to protect. The five intended outcomes of information privacy law thus change markedly and transit from protection of autonomy to situated inter-subjectivity, power vacuums to interruptions of modulation, transactional operation to boundary management, information asymmetries to social shaping and balancing mechanisms to exposing modulation.

At the heart of this reformulated movement is Cohen's work on modulated forms of power, which describes the consequences and challenges that arise from the collected world. Modulation offers a more complex frame through which to view the operation of power in the collected world, because Cohen considers power structurally across several different spectrums. These spectrums cover the broader political economy behind ever-cumulative forms of informational capitalism and its concomitant surveillant logics, infrastructural requirements and device designs – in other words, the use of modulated power flows from the macro structures of informational capitalism through to micro activities of sensorised data collections. This discussion therefore encapsulates many of the concerns raised in earlier parts of the book. More importantly, Cohen's work provides a platform that allows for new forms of information privacy law to develop that are explicitly designed for the power-related consequences of the collected world.

Chapter 9 examines how new forms of information privacy law could develop to interrupt modulated forms of power. It highlights some design points for future legal reform. The design points are not exhaustive, but they outline some key areas that would allow reforms to develop based on Cohen's principles of semantic discontinuity and operational accountability. The implementation of these principles would require some form of detachment from information privacy's core process protections, so that the law could apply in gaps and spaces at the outskirts of process. These gaps and spaces are important because, as Cohen argues, this is where selfhood flourishes and would therefore be a prime target for modulated forms of data collection. The design points would allow protection of gaps and spaces through the construction of new boundary options that create pauses in seamless forms of data collection and analysis. All of this would assist in information privacy law's new role in exposing modulation.

Chapter 9 also returns to Chapter 6's analysis and highlights the need for relational forms of regulated information and a collection principle based on fairness. The jurisdictional considerations introduced in Chapter 6 thus come to the fore. Chapter 9 contends that a greater focus on relational forms of personal information, such as personal data in the EU, would be an important interrupter of seamless application because it would have the effect of extending information privacy law's coverage from process to gaps and spaces. Such a focus would extend coverage from specific types of information to power-related activities involving data collection and exchange.

It is also clear that information privacy law protections are required at the point of data collection. Chapter 6 outlines the debate against collection protections based on the notice and consent model predominant in the United States. That is quite simply the wrong approach for the collected world. As the book demonstrates throughout its three parts, sensor data collections are crucial to understanding the challenges that arise from the collected world and the modulated implications that could flow. It is time to recognise that privacy policies simply do not work and that information privacy law therefore needs to consider reformulating existing protections or formulating new ones. The resulting analysis focuses on the former, but Chapter 9 also considers the latter. New legal vocabularies and new ways of incentivising value discourse, as well as compliance orientations, in data-collecting institutions are required.

1.2.4   A Smart, Collected or Modulated World?

The book concludes, in Chapter 10, by asking the question of whether we are heading for a smart, collected or modulated world. The smart world is an uncritical one, where the technological wonderment of sensorised devices and seamless service provision is accepted without question. The collected world provides a different view of the smart world, albeit largely descriptive – one that highlights the underlying logics at play in relation to ubiquitous collections of sensor data. The modulated world is defiantly critical and is intended to surface the normative basis of embedded power that flows in modulated forms of knowledge production and prescriptive governance.

It would be easy to deliberate on these considerations from a largely negative and dystopian perspective. But let us not do so.

The modulated world is not yet fully upon us, and whether it manifests in its most abominable form is in our hands. As Cohen beautifully points out, one of the simple ways we can create a world that ensures greater benefits from technological integration, including sensorised data collections, involves our unceasing ability to tinker and play. In that sense, consider this book as playful tinkering with the concept and application of information privacy law for a collected world.

Part I

The Collected World

2 The Smart World Is the Collected World

2.1 Introduction

The smart world appears to be taking shape around us. Our societies are increasingly populated with the smart devices that make up the 'Internet of Things'.[1] Sensorisation of environmental spaces is unfolding at a rapid rate as we develop smart buildings[2] such as the smart home, store or workplace. Sensorised networks and infrastructures now make our broader environments, such as the smart city, an unfolding reality.[3] The components of the smart world should provide significant benefits. Our smarter cities will be more resource efficient and safer places to live.[4] Our homes will understand our needs and tailor their resources more effectively to meet our demands.[5] Our personal devices will track our moods and behaviours to shape and work out our present and future needs.[6]

The onset of the smart world entails radical technological development which could furnish positive opportunities of better understanding individual and collective behaviour to better target service delivery.[7]

---

[1] Maciej Kranz, *Building the Internet of Things: Implement New Business Models, Disrupt Competitors, and Transform Your Industry* (Wiley 2017); Bruce Sinclair, *IoT Inc: How Your Company Can Use the Internet of Things to Win in the Outcome Economy* (McGraw-Hill Education 2017).

[2] Daniel Kellmereit and Daniel Obodovski, *The Silent Intelligence: The Internet of Things* (DND Ventures LLC 2013); Robert Scoble and Shel Israel, *Age of Context: Mobile, Sensors, Data and the Future of Privacy* (Patrick Brewster Press 2014).

[3] Anthony Townsend and Jeremy Arthur, *Smart Cities: Big Data, Civic Hackers, and the Quest for a New Utopia* (Audible Studios 2013); Stephen Goldsmith and Susan Crawford, *The Responsive City* (Jossey-Bass 2014).

[4] Mark Koh and Brian Koh, 'Sense and the City' (*Frost & Sullivan*, 30 July 2014) <www.frost.com/prod/servlet/cpo/291680637> accessed 20 June 2019.

[5] Francis K Aldrich, 'Smart Homes: Past and Present' in Richard Harper (ed), *Inside the Smart Home* (Springer 2003) 34–5.

[6] Sara Bannerman, 'Relational Privacy and the Networked Governance of the Self' [2018] Information, Communication & Society 1, 8.

[7] Erik Brynjolfsson and Andrew McAfee, *The Second Machine Age: Work, Progress, and Prosperity in a Time of Brilliant Technologies* (W. W. Norton & Company 2014).

This chapter describes these technological developments in greater depth to provide a platform for later discussion that highlights the endless drive towards ubiquitous data collection based on sensor-generated data. The smart world is outlined in three ways – namely, through smart individuals, smart buildings and smart environments.[8]

First, the chapter examines the development of the smart individual and our use of sensorised devices that are becoming embedded into our lives. The most perceptible device is the smartphone. The volume of sensors now packed into a smartphone mean that this device can be used for multiple purposes.[9] The smartphone is no longer a device that simply provides the means for verbal communication. It is now a map, a pedometer, a compass, a camera, a health monitor, a barometer, a thermometer, an earthquake monitor, an emotional guide, a music and entertainment centre, a television, a games console, a diary, a calendar, a timekeeper, a photo album, a guitar tuner and an intersection of multiple and different communication sources. In other words, the previous equivalent of many devices is now packed into one device that fits snugly into our pockets and our handbags. Furthermore, the advent of smart watches, smart glasses, smart clothes, even smart pills, means that our daily activities can be recorded in many ways.[10] Which takes us to the next step, the advent of the truly 'smart individual' and the use of sensorised body implants, where the device physically becomes part of the person.[11]

Second, the chapter examines the development of the 'smart building' and the increasing sensorisation of the day-to-day spaces whose sanctity we regularly take for granted – most notably the home, which will be explored further in Chapter 3. However, it is not just the home that is becoming increasingly smarter. Our working spaces, the devices we use and the working processes we operate within are all becoming increasingly sensorised to the extent that more data is generated about our working activities and behaviours.[12] From the moment we enter a working space, our activities may now be recorded, from switching on the

---

[8] This analysis of the smart world's components is similar to that of Meg Leta Jones, 'Privacy without Screens & the Internet of Other People's Things' 51 Idaho Law Review 639.

[9] David Nield, 'All the Sensors in Your Smartphone, and How They Work' (*Gizmodo*, 28 July 2017) <www.gizmodo.com.au/2017/07/all-the-sensors-in-your-smartphone-and-how-they-work/> accessed 20 June 2019.

[10] Marjolein Lanzing, 'The Transparent Self' (2016) 18 Ethics and Information Technology 9, 10.

[11] Luiz Costa, *Virtuality and Capabilities in a World of Ambient Intelligence: New Challenges to Privacy and Data Protection* (Springer 2016) 36.

[12] Mark Burdon and Paul Harpur, 'Re-conceptualising Privacy and Discrimination in an Age of Talent Analytics' (2014) 37 University of New South Wales Law Journal 679.

computer in the morning to turning off the lights at the end of the day.[13] The development of sensorised hard hats, machinery and industrial processes means that employees of all kinds now increasingly interact with smart workplaces.[14] The final space-based activity covered in the chapter is the development of the smart store. A significant number of store-based devices and services now monitor collective and individual shopping habits and attempt to provide an increasingly personalised service.[15]

Finally, the chapter examines the smart environment and the development of smart cities, smarter infrastructure and smart farms. The smart environment encapsulates the development of the smart individual and the smart building, to situate sensorised device development within the context of wider infrastructure. The concept of 'smartness' in this sense shifts from the provision of enhanced personalised services, by knowing more about individual behaviours and activities, to the provision of societal services such as safety, transport and the more efficient use of resources.[16] The considerations inherent in the smart environment shift towards connecting the smart individual with its companion smart building and wider environments.

The formulation of the smart world starts to take shape. It is a world in which everything becomes connected, either directly or tangentially. For this smart world to function, data about everything needs to be collected. Otherwise, the smart individual, building and environment cannot operate seamlessly together. The smart world is then a collected world. However, this point is often bypassed by the underlying heuristic frameworks of smartness such as the 'Internet of Things' and smart cities. These frameworks do little, if anything, to reveal the data collection requirements of the smart world.[17] Those data requirements involve collection by individual devices that flow through to aggregate collectives of data which are analysed to better understand individual and community patterns of behaviour.[18] Let's start with the device that has been pivotal to the development of the smart individual and its concomitant world: the smartphone.

---

[13] Thomas H Davenport, Jeanne Harris and Jeremy Shapiro, 'Competing on Talent Analytics' (2010) 88 Harvard Business Review 52.

[14] Aaron Morby, 'Smart Hard Hats Sound Alarm as Workers Tire' (*Construction Enquirer*, 2017) <www.constructionenquirer.com/2017/02/15/smart-hard-hats-sound-alarm-as-workers-tire/> accessed 20 June 2019.

[15] Joseph Turow, *The Aisles Have Eyes: How Retailers Track Your Shopping, Strip Your Privacy, and Define Your Power* (Yale University Press 2017).

[16] Lillian Edwards, 'Privacy, Security and Data Protection in Smart Cities: A Critical EU Law Perspective' (2016) 2 European Data Protection Law Review 28, 31.

[17] Rob Kitchin, 'The Real-Time City? Big Data and Smart Urbanism' (2014) 79 GeoJournal 1, 3–5.

[18] David Murakami Wood and Debra Mackinnon, 'Partial Platforms and Oligoptic Surveillance in the Smart City' (2019) 17 Surveillance & Society 176.

2.2   Smart Individuals

The development of the smart individual is intimately linked to the rise of smart devices, most notably the smartphone. For the purposes of this book, the smart individual is one who uses smart devices for conducting their daily life. Likewise, a smart device is a sensorised device, a device with inbuilt sensors, which has the capacity to sense and record individual activity to provide enhanced service delivery. The sensorised device can record finer data because of the quantity and quality of data collection that takes place.

In general terms, a sensor is a device that measures or detects an event or state and translates this measurement or detection into a signal.[19] As such, sensors do not monitor and listen per se. Rather, sensors detect and record, and they do not rely on direct and conscious registration on the part of those being sensed. Data collected by sensors requires a form of passive monitoring on behalf of the device user.[20] Sensor-generated collection of data, such as those conducted by smartphone, provide a realm of interactivity between users and data collectors in which users are rarely informed or aware of the data collection taking place.[21]

There are benefits to this lack of user awareness because it is claimed that data collected in this way provides a more accurate representation of the activities conducted by individuals. The smart individual can use their smart device to become aware of patterns of behaviour that they would otherwise not be able to identify or contemplate themselves. In other words, the smart device is sufficiently able to sense the activities of life that a 'data double' can be replicated and analysed to better effect.[22]

Before we consider some smartphone uses that demonstrate the positives of smart individualism, a brief overview of the smartphone is necessary. The rise of the smartphone has been exponential. Apple's iPhone was first released to the market in 2007. Since then, Apple has shipped nearly 1.5 billion iPhones worldwide.[23] It is estimated that the total number of smartphone users currently in operation across the

---

[19] Johan H Huijsing, 'Smart Sensor Systems: Why? Where? How?' in Gerard Meijer (ed), *Smart Sensor Systems* (John Wiley & Sons 2008) 3.

[20] Mark Andrejevic and Mark Burdon, 'Defining the Sensor Society' (2015) 16 Television & New Media 19, 24.

[21] Ibid 20.

[22] Deborah Lupton, 'Feeling Your Data: Touch and Making Sense of Personal Digital Data' (2017) 19 New Media & Society 1599, 1602.

[23] 'Unit Sales of the Apple iPhone Worldwide from 2007 to 2018' (*Statista*, 2019) <www.statista.com/statistics/276306/global-apple-iphone-sales-since-fiscal-year-2007/> accessed 20 June 2019.

globe is over 12.1 billion.[24] These statistics are important in the context of understanding the smart world because of the number of sensors packed into smartphones and what the devices can record.

The average smartphone contains up to a dozen different sensors which can sense and record a number of different activities.[25] A gyroscope detects when the phone is being rotated or twisted, and another sensor, an accelerometer, measures the degree of orientation change and adjusts the screen direction in relation to the angle it has been tilted.[26] Touch sensors record when force is being applied to the screen, which allows the user to swipe, scroll and choose applications via their smartphone interface. Proximity sensors detect when the phone is located close to our bodies, to preserve battery life and ensure that phone calls are not concluded accidentally on contact with the ear.[27] Specifically designed environmental sensors can also detect temperature and humidity within the phone's immediate locale.[28]

In addition to these specifically designed sensors, the smartphone has several production components that can be used for sensor related data collections. For example, the phone's Global Positioning System (GPS), wireless access card, Bluetooth chip and cellular connection features can be used for determining and recording location and geographical positioning details.[29] The phone's microphone can be used to sense sound and to automatically activate audio-recording capabilities,[30] while battery sensors detect the demands of ongoing usage in correspondence with battery life, in order to provide a warning for users that the phone's battery is running low.[31]

---

[24] 'Number of Smartphone Users Worldwide from 2014 to 2020' (*Statista*, 2015) <www.statista.com/statistics/330695/number-of-smartphone-users-worldwide/> accessed 20 June 2019.

[25] A Vaughn and others, 'Activity Detection and Analysis Using Smartphone Sensors' (2018 IEEE International Conference on Information Reuse and Integration (IRI), Salt Lake City, July 2018) 103.

[26] Ibid 104–5.

[27] 'Sensors – Definition' (*GSM Arena*, 2019) <www.gsmarena.com/glossary.php3?term=sensors> accessed 20 June 2019.

[28] 'Turn Your Smartphone into Any Kind of Sensor' (National Aeronautics and Space Administration, 24 June 2019) <www.nasa.gov/offices/oct/feature/turn-your-smartphone-into-any-kind-of-sensor> accessed 24 June 2019.

[29] David Nield, 'How Location Tracking Actually Works on Your Smartphone' (*Gizmodo*, 4 September 2018) <www.gizmodo.com.au/2018/09/how-location-tracking-actually-works-on-your-smartphone/> accessed 20 June 2019.

[30] Manisha Priyadarshini, 'Which Sensors Do I Have in My Smartphone? How Do They Work?' (*Fossbytes*, 25 September 2018) <https://fossbytes.com/which-smartphone-sensors-how-work/> accessed 20 June 2019.

[31] 'Sensors Overview' (Android Developers 2019) <https://developer.android.com/guide/topics/sensors/sensors_overview> accessed 20 June 2019.

Smartphone sensors, as just briefly inventoried, become important in the context of the smart individual because the device can now be used in a proliferation of different ways to sense, record and monitor the activities of users. The phone's sense-recording capabilities provide the impetus for data capture, which in turn can lead to new analytical insights regarding an individual's daily life. The sensory capacities of smart devices allow individuals to better 'quantify' their existence and thus make more informed, data-driven choices, based on a better understanding of their own behaviour.[32]

The smart individual is a product of the sensing apparatus provided by the ubiquitous accompaniment of smart devices. It is the device that enables the self-quantification processes necessary to reach the sensing enlightenment of individualised smartness. This device- and sensor-based enlightenment is producing some previously unquantifiable insights into non-smart life. In other words, our smartphones, in conjunction with specifically designed apps, are generating information about us that we normally could not perceive ourselves.

Smartphone voice recordings can now be used in a long-term process to predict future medical problems, such as the onset of a Parkinson's attack or signs of conditions like Alzheimer's.[33] Both Android and iOS operating systems now have inbuilt health-related monitoring capacities that utilise the phone's sensors to self-monitor body heat, heart rate and fitness levels. A smartphone can be used to identify better sleeping patterns thanks to its accelerometer.[34] The phone, in conjunction with apps such as Sleep Cycle,[35] can be used to sense and determine how much tossing and turning has taken place and at what point during the night.[36]

We are also on the cusp of smartphone sensor-based usage that can identify and predict emotional patterns. Microsoft, Apple, Samsung and Google have all submitted patents to register smartphone usage

[32] Gina Neff and Dawn Nafus, *Self-Tracking* (The MIT Press 2016) 7–8; Mark Hoogendoorn and Burkhardt Funk, *Machine Learning for the Quantified Self on the Art of Learning from Sensory Data* (Springer 2018) 1–2.

[33] Jacki Liddle and others, 'Measuring the Lifespace of People with Parkinson's Disease Using Smartphones: Proof of Principle' (2014) 16 Journal of Medical Internet Research e13.

[34] Z Chen and others, 'Unobtrusive Sleep Monitoring Using Smartphones' (Proceedings of the 2013 7th International Conference on Pervasive Computing Technologies for Healthcare and Workshops, PervasiveHealth, Venice, May 2013).

[35] 'Wake Up Easy with Sleep Cycle' (Sleep Cycle 2019) <www.sleepcycle.com/> accessed 20 June 2019.

[36] Dhvanesh Adhiya, 'Best Sleep Tracking iPhone Apps of 2019 for Optimal Sleep Cycles and Health' (*iGeeks Blog*, 16 February 2019) <www.igeeksblog.com/best-sleep-tracker-apps-for-iphone/> accessed 20 June 2019.

patterns that can be correlated to emotional states. Samsung researchers identified the basic emotions of a user – whether they were happy, sad, fearful, surprised or angry – from the way a smartphone was used. The use of a certain symbol in a text or how much the phone is shaking can also be used to partially determine the emotional state of the user.[37] Microsoft researchers identified the emotional patterns of Windows phone users from how the user operated the phone.[38] Even the on and off patterns of the touch screen and how apps are used, particularly in relation to communication patterns, could be correlated to reveal certain emotional behaviours.[39]

Smartphones are also now used for a range of different purposes, which means that data can now be collected about several different activities. The development of mobile payment services, such as Apple Pay, means that services and goods can be purchased through the use of a smartphone.[40] Technology developers are also looking at the next generation of passports, which will be provided through a smartphone.[41] It will no longer be necessary to show physical documentation to customs officials, and in fact we will not need any documentation at all, other than that built into our smartphone.[42] These extended uses of smartphones also have important implications regarding activity monitoring of individuals, because the phone is now being used as different objects. The smartphone as a payment device will allow the collection of data in relation to our payment patterns that can be accumulated over a period of time. The smartphone as a passport will provide a more fine-grained

[37] Duncan Graham-Rowe, 'A Smart Phone That Knows You're Angry' (*MIT Technology Review*) <www.technologyreview.com/s/426560/a-smart-phone-that-knows-youre-angry/> accessed 2 May 2019.

[38] Robert Likamwa and others, 'MoodScope: Building a Mood Sensor from Smartphone Usage Patterns' (Proceeding of the 11th Annual International Conference on Mobile Systems, Applications, and Services, MobiSys '13, Taipei, June 2013); Mike Elgan, 'Apple's iPhone X Proves It: Silicon Valley Is Getting Emotional' (*Computerworld*, 28 October 2017) <www.computerworld.com/article/3235424/apples-iphone-x-proves-it-silicon-valley-is-getting-emotional.html> accessed 20 June 2019.

[39] Xiao Zhang and others, 'MoodExplorer: Towards Compound Emotion Detection via Smartphone Sensing' (2018) [ACM] 1 Proceedings of the ACM on Interactive, Mobile, Wearable and Ubiquitous Technologies 1, 12.

[40] 'Apple Pay: Cashless Made Effortless' (Apple 2019) <www.apple.com/au/apple-pay/> accessed 20 June 2019.

[41] James Titcomb, 'Your Smartphone Could Be Your Next Passport' *The Telegraph* (London, 29 March 2016) Technology <www.telegraph.co.uk/technology/2016/03/29/your-smartphone-could-be-your-next-passport/> accessed 2 May 2019.

[42] Ibid. See the US Customs and Border Protection app, Mobile Passport Control, as a sign of development in the phone-as-passport space, 'Mobile Passport Control' (U.S. Customs and Border Protection, 12 March 2019) <www.cbp.gov/travel/us-citizens/mobile-passport-control> accessed 20 June 2019.

understanding of our movement behaviours in travel – and specifically at airports or other travel centres.

It is also not just different smartphone uses that enhance the possibilities for increased data collection. Smartphone apps are being equipped with an array of specialised external sensors. Tempdrop, for example, is a wearable body sensor that allows women to record their basal body temperature during sleep in order to track menstrual cycles and identify the times when a woman's body is most likely to be fertile.[43] External sensors are perhaps more familiar in the space of fitness tracking, where external sensors such as a Fitbit or a Jawbone watch are designed to track physical activity and record body movement reactions such as heartbeat rates. The smartphone is still intimately linked to this external sensor, as the sensor requires a smartphone app for the user to record and access fitness data.[44] The external sensor is an enhanced, sensorised prosthesis of the smartphone.[45] These sensors are becoming increasingly popular and are now being used to detect and measure a range of different activities, including optimum tennis swing and even the amount of spin on a ball.[46]

The use of external sensors connected to a smartphone leads into a different development – wearable technologies – in other words, sensor-implanted wearable objects such as glasses, watches or clothes. The most prominent and controversial of these wearable developments has been the now-infamous Google Glass,[47] a pair of smart glasses that could be used to take photos or videos, search the internet, get directions and read email, tweets, etc.[48] Like Google Glass, smart watches are wrist-sized sensorised computers that pack a range of sensors similar

---

[43] Catherine Shu, 'Tempdrop Is a Wearable Body Temperature Sensor for Fertility Tracking' (*TechCrunch*, 11 April 2014) <http://techcrunch.com/2014/04/11/tempdrop-is-a-wearable-body-temperature-sensor-for-fertility-tracking/> accessed 20 June 2019.

[44] 'Our Technology' (Fitbit 2019) <www.fitbit.com/au/technology> accessed 20 June 2019.

[45] Gavin JD Smith, 'Data Doxa: The Affective Consequences of Data Practices' (2018) 5 Big Data & Society 1, 11.

[46] 'Smart Tennis Sensor' (Sony 2016) <www.sony.com.au/microsite/tennis/> accessed 20 June 2019.

[47] Dana Schuster, 'The Revolt against Google "Glassholes"' *New York Post* (New York, 14 July 2014) Tech <https://nypost.com/2014/07/14/is-google-glass-cool-or-just-plain-creepy/>; Nick Bilton, 'Why Google Glass Broke' *The New York Times* (New York, 4 February 2015) Style, Disruptions <www.nytimes.com/2015/02/05/style/why-google-glass-broke.html>.

[48] Joanna Stern, 'Google Glass: What You Can and Can't Do with Google's Wearable Computer' (*ABC News*, 2 May 2013) <http://abcnews.go.com/Technology/google-glass-googles-wearable-gadget/story?id=19091948> accessed 20 June 2019.

to those found in a smartphone into the much smaller frame of a large watch. The smart watch is physically attached to our bodies, which means that it can reveal even more fine-grained data about our activities. For example, data collected from the motion sensors can be used to analyse patterns of movement that can even reveal the identification of PINs typed into a keypad.[49]

And it is not just watches that are becoming targets of smart development. Smart bracelets and smart rings can be connected to a smartphone and light up or vibrate when the smartphone receives a new tweet or notification that an Uber ride is close.[50] Similarly, Levi's Commuter Trucker jacket, a collaboration between Jacquard and Google, will buzz when the required Uber is close and also provide the name of the driver through the brush of a sleeve.[51] Sportswear is also a target of increasing sensorisation. Smart shoes monitor gait patterns and can suggest changes to walking and running movements to save running shoes from unnecessary wear and tear.[52] Ralph Lauren Polo shirts now have implanted biosensors that track the fitness buff's calorie-burning activity and the intensity of their movements.[53]

Biosensors and electronic impulses provide insight into a world of miniature sensors to record data about inner physical workings. Smart pills record infinitely more accurate data about an individual's body temperature and heart rate than conventional processes do. The smart pill is a battery-powered sensor that broadcasts data wirelessly to a designated wireless receiver.[54] A user swallows the pill, and the pill then transmits body temperature and other sensor-related data about

---

[49] Tony Beltramelli and Sebastian Risi, 'Deep-Spying: Spying Using Smartwatch and Deep Learning' (Cornell University 2015) <https://arxiv.org/abs/1512.05616> accessed 20 June 2019.

[50] Noelle Sciacca, 'Ringly Launches Fashionable Smart Bracelets' (*Mashable*, 29 March 2016) <http://mashable.com/2016/03/29/ringly-smart-bracelets/#o10IKEdeqPqV> accessed 20 June 2019.

[51] Clare Press, 'Wearable Tech and Smart Clothes Are Poised to Revolutionise Fashion' *Australian Financial Review* (Melbourne, 13 March 2019) Luxury <www.afr.com/brand/luxury/wearable-tech-and-smart-clothes-are-poised-to-revolutionise-fashion-20190220-h1bhse>.

[52] Kieran Alger, 'Connecting the Feet: The Future of Smart Running Shoes' (*Wareable*, 5 February 2018) <www.wareable.com/running/future-of-smart-running-shoes-340> accessed 20 June 2019.

[53] Hunter Atkins, 'Ralph Lauren's Futuristic Fashion: The PoloTech Smart Shirt' (*Forbes*, 8 September 2015) <www.forbes.com/sites/hunteratkins/2015/09/08/ralph-laurens-futuristic-fashion-the-polotech-smart-shirt/#2bb1793c59a5> accessed 20 June 2019.

[54] Amelia R Montgomery, 'Just What the Doctor Ordered: Protecting Privacy Without Impeding Development of Digital Pills' (2016) 19 Vanderbilt Journal of Entertainment and Technology Law 147, 151.

stomach content for further analysis.[55] The pill passes through the body's system within a 24- to 48-hour period.

This last sensorised development in our discussion is perhaps the most radical of all. If we consider the advent of the smartphone and the increasing miniaturisation of sensor-based devices that has followed, we can see that the relationship between sensors and humans is becoming increasingly more intimate.[56] Smartphones that travel with us in our daily lives identify and infer our behaviours. Those inferences become more fine-grained when the process of sensorisation is transferred to body-worn objects such as clothes, jewellery and watches. It becomes easier to identify the types of activities taking place, such as withdrawing money from an ATM, and the components of that activity, for example the identification of a bank-card PIN. Further miniaturisation, in the form of smart pills, reduces the spatial distance between human and sensor, and ever-finer measures of the human body become ascertainable. All of this leads to the ultimate and final step of device sensorisation and human activity – biosensors – where the physical and the technical are melded together.[57] In other words, the human and the sensor become one.

Humans of course have the capacity to sense. We know that a flame is hot, but we do not have the capability of a fine-grained sensor to determine the temperature of the flame. Humans as sensor-based data recorders are fallible because we cannot conduct an accurate measurement. That may well be about to change with the implantation of sensor devices into humans. Grindhouse Wetwear is a collective of biohackers that is developing the next generation of sensorised technology – the bio-implant.[58] This type of device includes a circuit board that is inserted into the arm which can detect a range of personal health measures such as blood glucose and oxygen levels, heart rates and blood pressure.[59] The sensorised examination of internal body chemical production can then be used to understand and predict stresses and emotions from a cell-level perspective. A sensor implant on the back

---

[55] Erin Brodwin, 'A Silicon Valley Company Just Launched "Smart" Cancer Pills That Track You with Tiny Sensors Stamped into Your Medications' (*Business Insider*, 18 January 2019) <www.businessinsider.com.au/digital-smart-pill-drug-sensor-company-expands-cancer-proteus-2019-1?r=US&IR=T> accessed 20 June 2019.

[56] Lupton 1601.

[57] George K Knopf and Amarjeet S Bassi, 'Introduction to Biosensors and Bioelectronics' in George K Knopf and Amarjeet S Bassi (eds), *Smart Biosensor Technology* (2nd edn, CRC Press 2019) 3–4.

[58] Jacob Roberts, 'Hacking Humans' *Distillations* (Philadelphia, 8 March 2017) Tools & Technology <www.sciencehistory.org/distillations/hacking-humans>.

[59] Ibid.

of the hand can control other smart devices and objects by touch, thus embedding device control within the human form.

We are not yet at a point in time where we must seriously consider the ubiquitous consequences of most human bodies as specific data collection sensors. However, given the widespread use of smartphones and the vastly increasing markets for other sensor-based personal devices, it is becoming essential to consider the implications of life in a smart world. Such considerations become even more pressing when the context of the smart world is extended beyond devices that are oriented towards the individual, to the next level – the advent of smart buildings and the development of sensorised spaces.

2.3   Smart Buildings

Given the proliferation of smart devices for an increasing number of individual uses, it should be no surprise that the reach of sensorisation has permeated beyond the tracking of individual activities to the monitoring of physical spaces. However, unlike the device-based sensing that takes place in the realm of the smart individual, smart building sensing derives from a number of different sensorised devices and is not specifically targeted towards a better understanding of the behavioural patterns and activities of the individual per se. Rather, it is about a better understanding of what happens in the space itself and how that understanding can be used to inform more efficient and effective uses of the space and the devices operating within it.[60] There is an overlap between spatial monitoring and individual tracking, as the former will necessarily involve the sensing of all activities conducted in the space, including human activities. The smart building will not be truly smart if it cannot account for the requirements, actions and desires of its human inhabitants. As such, there is still an intimate and direct link between the agency of smart individuals and the sensing apparatus of the smart building. This is most clear in our most cherished space, the home, which is covered extensively in Chapter 3.

Smart homes are not the only smart buildings. The smart workplace is also becoming increasingly sensorised to gain a better understanding of work-based processes and activities. The tools of the everyday office – the photocopier, the computer, the phone, etc. – become de facto sensors and produce metadata about their use.[61] Metadata is data

---

[60] Kitchin, 'The Real-Time City?' 4.
[61] Robert Sprague, 'Welcome to the Machine: Privacy and Workplace Implications of Predictive Analytics' (2015) 21 Richmond Journal of Law & Technology 1, 34.

about data, and in the context of the smart workplace, metadata generation refers to automated data produced by internal information systems
regarding system use by employees. These metadata 'breadcrumbs'
provide a treasure trove of data about internal processes, because
they essentially compose a log of an employee's activity throughout
the working day.[62] Audit trail logs record when an employee uses a
particular device or database. Keystroke logging software records the
keystroke movements on any given organisational keyboard. Swipe
cards and card readers provide data on when an individual accesses or
departs a secured office door. Telephone records detail the times and
locations of phone calls. Even data from photocopiers can be used if
configured correctly.[63]

Specifically designed sensors are also starting to come to light. The
sociometer[64] was designed by the MIT Human Dynamics laboratory
in the early 2000s and is a small device, similar to an identification
card, worn by employees that consists of a number of different sensors.[65] These sensors provide different data measurements – such as
location, sound and motion – and it thus becomes possible to monitor
the detailed activities of employees and their day-to-day interactions
to provide a fine-grained insight into employee and organisational
activity.[66] Workspaces themselves are also increasingly sensorised to
ascertain optimal occupancy levels. OccupEye is an intelligent space-
monitoring sensor that records when an individual employee is at
their desk or the activity of a designated space such as a meeting
room.[67]

The results of sensorisation are producing new insights and significant financial benefits. Language patterns in email messages – as

---

[62] Ben Waber, *People Analytics: How Social Sensing Technology Will Transform Business and What It Tells Us about the Future of Work* (FT Press 2013) 7.

[63] Omer Tene and Jules Polonetsky, 'Big Data for All: Privacy and User Control in the Age of Analytics' (2013) 11 Northwestern Journal of Technology and Intellectual Property 240, 247–50.

[64] Waber 10.

[65] Tanzeem Choudhury and Sandy Pentland, 'Sensing and Modeling Human Networks Using the Sociometer' (Seventh IEEE International Symposium on Wearable Computers, ISWC 2003, October 2003, White Plains, New York).

[66] John Hall, 'Is Your Boss Watching You? Surveillance Device Tracks Employees' Movements in the Office, Sends Details of Conversations and Even Times Their Toilet Breaks' *Daily Mail* (London, 6 February 2014) Science <www.dailymail.co.uk/sciencetech/article-2552858/Workplace-surveillance-device-tracks-employees-movements-office-sending-boss-details-conversations-colleagues-long-time-spend-toilet.html>.

[67] 'A Typical Scenario' (OccupEye 2016) <www.occupeye.com/how-it-works-2/> accessed 20 June 2019.

opposed to the content of such messages – now provide reliable indicators of unhappy star performers,[68] as do increasing numbers of visits to LinkedIn or Facebook.[69] Even employee smiles are being captured by cameras to ascertain the effect on customers.[70] Those employees who socialise more frequently at the water cooler or in the office kitchen are not the slackers they were once thought to be. Instead, these are the workers who are more likely to stay with an employer because they identify their employment environment as a place that provides cohesive community support, a key factor in employee retention in some industrial sectors.[71]

Significant financial savings are also reported from increased sensorisation of the workplace. The international courier firm UPS installed telematic sensors to more than 46,000 delivery trucks that deliver over 16 million packages per day for over 8 million customers.[72] The sensors provide enhanced data on truck speed, directions taken by drivers and braking intensity. From the data, UPS was able to maximise driver performance by changing routes, which reduced its worldwide daily delivery schedule by 85 million miles. The cost savings were considerable, as UPS estimated that saving one daily mile driven by one driver saves the company overall $US 30 million.[73]

The UPS example shows that it is not just the office workspace that is becoming sensorised. The advent of the smart 'hard hat' and the sensorisation of construction machinery mean that construction sites are also becoming smart. The international construction firm Laing O'Rourke has developed a smart hard hat to proactively monitor individual employee workplace health and safety. The hard hat has several different sensors, including a sweatband sensor, a GPS tracker and an accelerometer that can monitor an employee's body temperature and heart rate to ensure their well-being.[74] Similarly, the Daqri hard hat employs a number of video and audio sensors that can be used to

[68] Don Peck, 'They're Watching You at Work' *The Atlantic* (Boston, 20 November 2013) Business <www.theatlantic.com/magazine/archive/2013/12/theyre-watching-you-at-work/354681/>.

[69] Ibid.

[70] Ibid.

[71] Waber 85.

[72] Thomas H Davenport and Jill Dyche, *Big Data in Big Companies* (International Institute for Analytics 2013) 4.

[73] Ibid 4.

[74] Aimee Chanthadavong, 'Laing O'Rourke Monitors Workers' Safety with a Smart Hardhat' (*ZDNet*, 28 November 2015) <www.zdnet.com/article/laing-orourke-monitors-workers-safety-with-a-smart-hardhat/> accessed 20 June 2019.

provide an enhanced reality of what the hard hat wearer is currently able to detect, thus assisting them to complete their work.[75]

The final building space considered is the smart store. Most of us must shop, whether it be for the weekly groceries or for more infrequent purchases such as clothing, other consumables or presents. There is value for stores in knowing more about the shopping patterns and behaviours of customers, and the smartphone is once again at the heart of sensorised developments.[76] A number of stores and malls now provide free Wi-Fi for customers. This is not just an altruistic gesture or one that is designed to attract the download-conscious shopper. Instead, the in-store use of Wi-Fi is a way to monitor the shopping activities of individuals in the smart store. If an individual connects to the store's Wi-Fi, the store can then track the footpaths taken by the shopper, whilst in the store and even beyond, depending on the strength of the wireless signal.[77] This innovative use of Wi-Fi has opened a whole new range of marketing strategies based on greater knowledge of personalised shopping habits.[78]

Another common form of smart store sensorisation is Bluetooth low-energy beacons. Wireless can monitor the activities of shoppers across a wider area – up to fifty metres or more – whereas beacons are more fine-grained and operate within a shorter distance.[79] A beacon is a Bluetooth signal receiver that can detect when a Bluetooth-active smartphone is in the vicinity.[80] Normally, the beacon is attached underneath a shelf or inconspicuously on a wall. They do not require the shopper to have logged into the smart store's Wi-Fi offering before they operate; any shopper who has Bluetooth activated on their phone can be monitored by beacons. It then becomes possible to get a much closer understanding of customer behaviour in relation to individual items. A beacon can ascertain how long a smartphone was in a specific area, thus leading to an inference of customer likes or dislikes,[81] including

[75] Aliya Barnnewll, 'The Daqri Smart Helmet Turns a Workforce into Robocops' (*Digital Trends*, 21 April 2015) <www.digitaltrends.com/cool-tech/daqri-smart-helmet-reads-world-around/> accessed 20 June 2019.

[76] Turow 4.

[77] Laurenz Schauer, 'Analyzing the Digital Society by Tracking Mobile Customer Devices' in Claudia Linnhoff-Popien, Ralf Schneider and Michael Zaddach (eds), *Digital Marketplaces Unleashed* (Springer 2017) 474–6.

[78] Turow 179.

[79] Patrick Dickinson and others, 'Indoor Positioning of Shoppers Using a Network of Bluetooth Low Energy Beacons' (2016 International Conference on Indoor Positioning and Indoor Navigation (IPIN), October 2016, Alcalá de Henares, Spain).

[80] Matt Silk, 'Beacons Are Not the New Black' 4 Journal of Retail Analytics 6, 7.

[81] Linda Bustos, 'Digital Retail: How In-Store Analytics Align with Online' (*GetElastic*, 25 June 2014) <www.getelastic.com/digital-retail-how-in-store-analytics-align-with-online/> accessed 20 June 2019.

whether a customer is clearly interested in a product but has decided not to purchase it.[82]

These are not the only sensorised technologies of the smart store. Thermal imaging detects body heat emissions of customers to create heat maps of popular customer locations, which can be used to determine the popularity of store items.[83] Infrared beams in doorways detect how many customers enter and exit the store. Radio-frequency identification (RFID) tags used in smart shelves indicate when the shelf is running low on stock.[84] Even 3D imagery is used. High-resolution cameras capture and create a three-dimensional view of an object, either an individual customer or a shopping cart, for an extended period of time.[85] The smart store can therefore track the activities of the customer all the way down to what is actually placed in the shopping basket.[86]

The most surprising development is the smart mannequin. Its beacons sense when a customer is in the immediate vicinity[87] and then connect to the customer's smartphone and track the location of that phone and even send messages to it. If a customer pauses in a spot close to the mannequin, an automated advert can be sent to the customer's smartphone that provides information about clothing located near the customer or the clothing worn by the mannequin.[88] The smart mannequin also employs hidden video cameras that are installed in its eyes. The video footage is used, in combination with facial recognition software, to identify the demographic characteristics of customers. An advert can then be sent to a certain type of customer – based on gender or age, for example – using the beacon transmitter.[89] The smart mannequin is no longer just a receptacle of clothing; it is now a customer tracking and communication device.

[82] 'PHeat: Heatmap Analytic' (RetailFlux 2019) <www.retailflux.com/heat-map/> accessed 20 June 2019.
[83] Mackenzie Lane, 'Location-Based Analytics Yield Customer and Inventory Insights' 4 Journal of Retail Analytics 21.
[84] Ibid.
[85] Nils Magne Larsen, Valdimar Sigurdsson and Jørgen Breivik, 'The Use of Observational Technology to Study In-Store Behavior: Consumer Choice, Video Surveillance, and Retail Analytics' (2017) 40 The Behavior Analyst 343, 354.
[86] Emmett Cox, Retail Analytics: The Secret Weapon (Wiley 2012) 16–17.
[87] 'Smart Beacon and Smart Mannequin' (Almax 2016) <www.almax-italy.com/uploaded/attachments/folder_split2-3.pdf> accessed 20 June 2019.
[88] Sean Poulter, 'Mannequin That Tells You What It's Wearing: Dummies That Talk to Shoppers via Their Smartphones Set to Be Switched on Today' Daily Mail (London, 12 August 2014) <www.dailymail.co.uk/news/article-2722399/Mannequin-tells-s-wearing-Dummies-talk-shoppers-smartphones-set-switched-today.html>.
[89] Almax, 'Smart Beacon and Smart Mannequin'.

These different aspects of the smart building, smart home, work-place and store lead into this chapter's final consideration of the smart world. Smart individuals live and work in smart buildings, and those buildings reside in significantly smarter environments. The focus here is on the advent of the smart city and its infrastructural components.

2.4   Smart Environments

Thus far, the focus has been on the sensorisation of personal devices and the effect on the smart individual and the sensorisation of spaces with the advent of the smart building. Attention now shifts from devices, and their uses in certain spaces, to networks of sensors and the societal considerations that arise from knowing more about how we live collectively. The smart environment connects the smart individual and life in smart buildings to the wider requirement of resourcing and how we live together as a society. This connection is most visible in the upcoming rush to smarten our cities and urban environments.

The pressures on city lives are becoming demonstrable. Cities are now facing voluble population pressures and higher levels of taxpayer demand. The burden is on city managers to provide a service for an ever-greater number of individuals at an ever-higher level of quality.[90] All of this, of course, comes at a time when fiscal resources are increasingly squeezed. Technology is seen as the saviour by many city managers. The rapid uptake in sensorised networks provides the opportunity for a much greater knowledge about how a city operates whilst providing a platform for more effective, streamlined and seamless service delivery.[91] The dawn of the smart city has been a notable feature of the past decade and has been driven by some of the largest technology companies.[92]

[90] Rob Kitchin, 'The Promise and Perils of Smart Cities' (*Society for Computers and Law*, 2016) <www.scl.org/articles/3385-the-promise-and-perils-of-smart-cities> accessed 20 June 2019.

[91] Claire Thorne and Catherine Griffiths, 'Smart, Smarter, Smartest: Redefining Our Cities' in Paola Renata Dameri and Camille Rosenthal-Sabroux (eds), *Smart City: How to Create Public and Economic Value with High Technology in Urban Space* (Springer International Publishing 2014) 90.

[92] Annalisa Cocchia, 'Smart and Digital City: A Systematic Literature Review' in Paola Renata Dameri and Camille Rosenthal-Sabroux (eds), *Smart City: How to Create Public and Economic Value with High Technology in Urban Space* (Springer International Publishing 2014) 26. This has also been a significant bone of contention about the role of private sector corporations in the provision of data collection and analysis for the delivery of public services: Kitchin, 'The Real-Time City?' (2014); Robert G Hollands, 'Critical Interventions into the Corporate Smart City' (2015) 8 Cambridge Journal of Regions, Economy and Society 61; Oscar H Gandy and Selena Nemorin, 'Toward a Political Economy of Nudge: Smart City Variations' [2018] Information, Communication & Society 1.

IBM is one of the tech leaders at the forefront of smart city development.[93] Cities, according to IBM, are complex infrastructures of multifaceted, interconnected and competing systems of service.[94] Success in service delivery, it is argued, is dependent on the effectiveness and efficiency of delivery in areas such as water and energy supply, communication networks, transport networks, public safety, education, business planning and internal operational structures.[95] The pressures facing cities are such that 'the business as usual' model of service delivery is no longer sufficient. Cities have to transform their systems, and their 'systems of systems', to make them more digitised, more interconnected and thus more intelligent.[96] Intelligence means sensorisation,[97] and sensor-generated data can assist the smart city to predict citizen behaviours which can be modelled into real-world change.[98]

The sensor, the data generated and the insight created are thus equal features of the smart individual, the smart building and the smart environment – except that in the latter the scale of application is much greater. As detailed earlier, the sheer number of sensors packed into a smartphone mean that the phone can be used to record the activities of individuals in many ways. This chapter has thus far focused on sensor collection for the benefit of individuals, but the sensitivity of sensor smartphone collections can also be used to benefit societies. A committed populous who are willing to share data in relation to smartphone or other sensorised device usage can create new ground-up sensor networks that collect data on societal functions and issues.[99] The individual smartphone or device consequently becomes part of a wider environmental network.

The MyShake app[100] utilises the smartphone's accelerometer to detect movement caused by an earthquake event. The app is designed

---

[93] David Murakami Wood, 'Smart City, Surveillance City' (*Society for Computers and Law*, 2015) <www.scl.org/articles/3405-smart-city-surveillance-city> accessed 20 June 2019.
[94] Susanne Dirks and Mary Keeling, *A Vision of Smarter Cities: How Cities Can Lead the Way into a Prosperous and Sustainable Future* (IBM Global Business Services 2009) 9.
[95] Ibid 11.
[96] Ibid 2.
[97] Charith Perera and others, 'Sensing as a Service Model for Smart Cities Supported by Internet of Things' (2013) 25 Transactions on Emerging Telecommunications Technologies 81.
[98] Thorne and Griffiths 92.
[99] Jennifer Gabrys, 'Citizen Sensing, Air Pollution and Fracking: From "Caring About Your Air" to Speculative Practices of Evidencing Harm' (2017) 65 The Sociological Review 172.
[100] 'MyShake' (UC Berkeley Seismology Lab 2019) <https://myshake.berkeley.edu/> accessed 20 June 2019.

to detect ordinary smartphone movements, such as when it moves in a user's pocket or handbag, and it can also detect the type of unique vibrations found in earthquakes.[101] Each smartphone with the MyShake app can send a warning message, and the accumulation of user readings and responses records earthquake activity across a given geographical location.[102] The NoiseTube[103] app detects noise pollution levels by recording sound levels across GPS locations which are aggregated to produce a heat map of city noise levels.[104] The first smartphone with air pollution sensors have also been trialled to test their ability to collectively monitor air pollution conditions based on individual smartphone geographical use.[105]

Individual smartphones are not the only sensor networks deployed in the smart city, as the city itself becomes a vast array of sensorisation.[106] As with the smart building, the smart city monitors its own processes. In doing so, the smart city provides the opportunity for better understanding urban life through the expansion of different networks that can monitor individual and collective activities. The essential service networks of the smart city become sensor networks that record and monitor the life that transpires around them.[107] The supply of essential services can thus provide the means for city-scale monitoring.

The city of Santander, in northern Spain, installed a dedicated wireless sensor network across the city. Up to sixty different sensors were packed into sensor nodes installed at various locations throughout the city. More than 12,000 fixed and mobile sensors were installed across a range of different objects and places.[108] The sensors were primarily

---

[101] Rosanna Xia and Rong-Gong Lin, 'Scientists Develop New App That Uses Your Cellphone to Detect Earthquakes' *Los Angeles Times* (Los Angeles, 12 February 2016) Local <www.latimes.com/local/lanow/la-me-ln-app-mobile-phone-detect-earthquakes-20160212-story.html> accessed 3 May 2019.

[102] Ibid.

[103] 'Turn Your Mobile Phone into an Environmental Sensor and Participate in the Monitoring of Noise Pollution' (*NoiseTube*, 2018) <www.noisetube.net/index.html#&panel1-1> accessed 20 June 2019.

[104] Elizabeth Rust, 'Map Some Noise: How Your Smartphone Can Help Tackle City Sound Pollution' *The Guardian* (London, 12 September 2012) Cities <www.theguardian.com/cities/2014/sep/12/map-noise-how-smartphone-help-tackle-city-sound-pollution-noisetube> accessed 1 May 2019.

[105] Mawutorli Nyarku and others, 'Mobile Phones as Monitors of Personal Exposure to Air Pollution: Is This the Future?' (2018) 13 PLoS One e0193150.

[106] Kitchin, 'The Real-Time City?' 4.

[107] OECD, *Smart Sensor Networks: Technologies and Applications for Green Growth* (OECD Digital Economy Papers 2009) 31–4.

[108] Janine S Hiller and Jordan M Blanke, 'Smart Cities, Big Data, and the Resilience of Privacy' (2017) 68 Hastings Law Journal 309, 317.

used for parking and lighting services but were also used to monitor air quality, radiation levels, temperature and noise levels.[109] In early 2014, the *New York Times* reported the installation of 171 sensor-connected LED fixtures at Newark Airport that serve as the backbone of a wide-ranging and intricate airport security system. The fixtures are part of an array of sensors and surveillance cameras that collect data to identify suspicious activities and patterns.[110] Similarly, in Las Vegas, smart street light fixtures have been implemented to monitor air pollution, track foot traffic and undertake surveillance in relation to law enforcement purposes.[111]

Thus far the focus has been on public sector requirements for governance of the smart city. Essential services are also delivered by a range of private sector actors.[112] The sensorisation of the smart city entails a number of different public and private sector bodies, most notably in the area of electricity provision and the development of smart grids.[113] 'Smart grid' generally refers to the sensorisation of electricity networks to deliver power more efficiently and effectively to smart buildings.[114] A vast system of different sensors enables two-way communication between the millions of different devices involved in the delivery of electricity. It therefore becomes possible to better calibrate such delivery across the network and to better meet consumer demand.[115]

The smart grid operates in conjunction with smart meters, which are the next generation of gas and electricity meters. Smart meters provide a few benefits for both user and supplier alike, because they generate near to real-time data on energy consumption.[116] For the supplier, the collective use of smart grids provides a much more

---

[109] Alberto Beilsa, 'The Smart City Project in Santander' (*Sensors Online*, 1 March 2013) <www.sensorsmag.com/wireless-applications/smart-city-project-santander-11152> accessed 20 June 2019.

[110] Diane Cardwell, 'At Newark Airport, the Lights Are On, and They're Watching You' *New York Times* (New York, 17 February 201) Business <www.nytimes.com/2014/02/18/business/at-newark-airport-the-lights-are-on-and-theyre-watching-you.html>; Perera and others.

[111] Tod Newcombe, 'The Rise of the Sensor-Based Smart City' *Government Technology* <www.govtech.com/data/The-Rise-of-the-Sensor-Based-City.html>.

[112] Edwards 32.

[113] OECD, *Emerging Issues: The Internet of Things* (Digital Economy Outlook 2015) 253.

[114] OECD, *Smart Sensor Networks*, 14.

[115] 'Grid Modernization and the Smart Grid' (*Office of Electricity Delivery* 2016) <www.energy.gov/oe/activities/technology-development/grid-modernization-and-smart-grid> accessed 20 June 2019.

[116] OECD, *Machine-to-Machine Communications: Connecting Billions of Devices* (OECD Digital Economy Papers 2012) 13.

detailed understanding of electricity usage at every stage in the grid. Demand can be identified much more quickly across the grid and in the home. The activities of the individual, the building and the environment are again connected, and it becomes possible to see, for the first time, the visible effects of individual action in the smart home and its concomitant impact across the grid. The smart individual's energy-consuming patterns can be identified by the supplier, the smart house and ultimately by the individuals themselves. The effect, for instance, of having all fifty Phillips Hue light bulbs switched on at the same time can be identified, and data can be relayed back to the user to consider whether such resource-depleting action is appropriate.

The smart city has also extended into pastures greener. It is not just cities that are becoming smarter. It is also rural areas.[117] Take, for example, the Sense-T project, which involves the sensorisation of the island of Tasmania, Australia, across several different agricultural sectors.[118] A vast array of sensors has been installed to measure temperature, solar radiation, soil moisture and other environmental factors. Some sensors even measure the heartbeats of farmed oysters.[119] Other sensor networks monitor water quality, individual farmer irrigation and potato storage to identify optimum storage conditions. Livestock also have implanted sensors, to better monitor animal health and pasture prediction.[120]

2.5   Conclusion: Smartness Means Data Collection

The utopian smart world is taking shape around us.[121] The smartness of personal devices means that we, as individuals, become smarter. The ever-increasing ability of sensors to detect and record the minutiae of life means that we can quantify our behaviours and patterns in ways in which we could not do so previously. Device detection makes us smarter, and individuals can respond to the enhanced knowledge provided by sensing apparatuses to improve their lives. In having devices that understand people better, smart buildings adjust to better meet

---

[117] OECD, *Smart Sensor Networks*, 37–8.
[118] 'Home' (Sense-T 2016) <www.sense-t.org.au/> accessed 20 June 2019.
[119] 'Data Driven Transformation' (Sense-T 2016) <www.sense-t.org.au/__data/assets/pdf_file/0009/844272/Oyster-Impact-Example.pdf> accessed 20 June 2019.
[120] 'Agriculture' (Sense-T 2016) <www.sense-t.org.au/projects-and-research/agriculture> accessed 20 June 2019.
[121] Murakami Wood.

individual needs – whilst at the same time serving societal goals of efficiently using resources. The smart home knows when to dim the smart light bulb based on our behavioural patterns and predetermined levels of efficient energy usage. The smart city is better able to understand the life that unfolds in its confines thanks to the sensorised activities of individuals and the sensorisation of infrastructure. The smart individual thus receives the full benefit of appropriate resource allocation. Under this idealised vision, everyone benefits from having greater knowledge about everything.

For the smart world to truly function effectively, there must be interconnection between human agency and the objects and infrastructures of life that support human activity. Smartness requires an interconnection between the individual, the building or object of activity and the infrastructural resources that enable both the interconnection and the activity. Once these different levels are connected, it then becomes possible for the smart world to truly take shape.

The smart individual is no longer a singular human entity surrounded by sensing devices. The smart home no longer stands alone on its block of land. The smart city is no longer populated by unknown individuals who are too infinitesimal to truly understand. Instead, the smart individual can become an earthquake detector or a noise pollution monitor for the city. At the same time, the smart city becomes a safer place through a better understanding of the population and therefore the activities that flow through it, including in the home.

The downside to this vision of connected smartness is vast forms of data collection and acquisition. The smart world cannot be smart without data on how the individual, the home and the city all operate. Smartness in the context of the smart world does not just entail the collection of data from singular or multiple entities or objects; it requires the collection of data from and about everything. By its nature, the smart world cannot be smart until data about everything is collected. Previously, such ideas would have been for the confines of science fiction. However, the radical reduction in the cost of sensor production, combined with the vast increase of data storage capacities, now means that the fantasy of recording data about everything is showing the likelihood of fruition. Consider Google Street View as an example.

The idea that the world's locations could be photographed, mapped and then published in a publicly available format would have been unthinkable thirty years ago. Yet here it is, and the data acquisition capacity of Google Street View is mindboggling. In 2012, it was estimated that the complete Google Maps project – which combines satellite, aerial and Street View imagery – amassed more than 20 petabytes of data, the equivalent of 21 million gigabytes or over 20,000 terabytes.[122] To put this in context, it is estimated that 1 petabyte would be enough to store the DNA profiles of the population of the United States[123] and that there would still be leftover space for those important photos and music.

The sheer volume of data that is now being collected is staggering. IBM claimed in 2013, for example, that every day about 2.5 quintillion bytes of data is generated, the data equivalent of a quarter million Libraries of Congress, and that 90 per cent of the world's stored data has been created in the past ten years.[124] In 2012, it was reported that Facebook alone reportedly entered 500 terabytes, the equivalent of about fifty Libraries of Congress, into its databases each day.[125] Databases continually fill up with data that is generated mechanically and automatically by a burgeoning array of sensors.[126] Nor is data generation and collection going to decrease.

This avalanche of data is providing the basis for the smart world, and the predictions of the past three decades regarding the onset of rapid and continuous data growth have, by and large, turned out to be accurate. We are now at a stage where the Internet has grown beyond past communication infrastructural capacities.[127] Digital storage has become more cost effective, allowing for the capture and storage of an ever-growing range of data.[128] The increased capture of data from

---

[122] Matt Petronzio, '11 Fascinating Facts about Google Maps' (*Mashable*, 2012) <http://mashable.com/2012/08/22/google-maps-facts/#yq3n_7UtUkqbwww.sense-t.org.au/projects-and-research/agriculture> accessed 20 June 2019.

[123] Brian McKenna, 'What Does a Petabyte Look Like?' (*Computer Weekly*, 2013) <www.computerweekly.com/feature/What-does-a-petabyte-look-like> accessed 20 June 2019.

[124] Ralph Jacobson, '2.5 Quintillion Bytes of Data Created Every Day' (*IBM*, 24 April 2013) <www.ibm.com/blogs/insights-on-business/consumer-products/2-5-quintillion-bytes-of-data-created-every-day-how-does-cpg-retail-manage-it/> accessed 20 June 2019.

[125] Eliza Kern, 'Facebook Is Collecting Your Data – 500 Terabytes a Day' (*GigaOm*, 22 August 2012) <http://gigaom.com/2012/08/22/facebook-is-collecting-your-data-500-terabytes-a-day/> accessed 20 June 2019.

[126] Bernard Marr, 'How Much Data Do We Create Every Day? The Mind-Blowing Stats Everyone Should Read' (*Forbes*, 21 May 2018) <www.forbes.com/sites/bernardmarr/2018/05/21/how-much-data-do-we-create-every-day-the-mind-blowing-stats-everyone-should-read/#34ba63d060ba> accessed 20 June 2019.

[127] Kerry Coffman and Andrew Odlyzko, 'The Size and Growth of the Internet' (1998) 3 First Monday.

[128] Peter Lyman and Hal Varian, 'How Much Information?' (2000) 6 JEP: The Journal of Electronic Publishing.

a variety of automated processes[129] has led to the development of new analytic processes[130] and new ways of visualising the meaning of findings generated from vast amounts of data.[131]

These predicted expansions also provide a stimulus for several significant claims and justifications for the use of collected data in the smart world. We are now seeing some fundamental re-constitutions for data collection, including a reconceptualisation of personal information as the 'new oil':[132] a new asset class and a critical source for corporate and governmental innovation and value. At the same time, it is also argued that the collection of data about everything, and the analytical processes to make sense of that data, can assist with policy implementation and international development by providing hitherto-unseen insights.[133]

So what does the collection of data about everything actually entail in the smart world? Unfortunately, the current and dominant heuristic frameworks for understanding these complex questions do not greatly assist us. The Internet of Things quite rightly refers to the proliferation of smart devices, highlighted earlier, and the concomitant interconnection with other devices, individual smartphones and the control systems of the smart home. However, the 'things' that this heuristic portrays are not the sensory-deprived objects that the use of the word 'thing' would suggest.[134] The smart devices, the things, as we have already seen, are anything but sense-less. They are in fact the opposite. They are sense packed and are capable of recording and monitoring a multitude of activities. The Internet of Things does not acknowledge the data collection requirements of the smart world. These things are really instruments of sensorised data collection that enable the onset of the smartness required for a smart world.

Behind the high-gloss façade of the smart world are infrastructures of data collection, acquisition and sense-making. If the smart world is indeed the collected world, then it is important to turn attention to the

---

[129] Bret Swanson and George Gilder, *Estimating the Exaflood: The Impact of Video and Rich Media on the Internet* (Discovery Institute 2008).
[130] Peter J Denning, 'Saving All the Bits' (1990) 78 American Scientist 402.
[131] Steve Bryson and others, 'Visually Exploring Gigabyte Data Sets in Real Time' (1999) 42 Communications of the ACM 82.
[132] World Economic Forum, *Personal Data: The Emergence of a New Asset Class* (2011) 5.
[133] World Economic Forum, *Big Data, Big Impact: New Possibilities for International Development* (2012) 4.
[134] Andrew Guthrie Ferguson, 'The Internet of Things and the Fourth Amendment of Effects' (2016) 104 California Law Review 805, 859.

embedded infrastructures that underpin and enable smartness to better understand the processes and justifications for a collected world. The remainder of Part I considers the collected world in greater depth and the constituent processes of sense-making, which demand the collection of more and more sensor data. The business logics fuelling ubiquitous collection and the ultimate end goal of the collected world are therefore intimately linked. Nowhere is this more clear than in the sensor data collection and communication anatomy of the smart home, as examined in Chapter 4.

3   The Smart Home: A Collected Target

3.1   Introduction

Chapter 2 introduced the notion of a collected world. It highlighted that the points of connection which enable smartness also require ubiquitous forms of data collection, particularly from sensorised devices and infrastructures. In this chapter, the key collection consequences of the collected world are examined through one of its prime commercial targets, the smart home. The home is a treasured and protected site of liberalism[1] recognised legally as a space for the autonomous growth of individuality.[2] It is a space sheltered from collective interference so that we can be our most natural. It is where we develop our most intimate relationships. The home is thus the citadel of the private domain.[3] It facilitates the historical accumulation of personal activities and allows for the facilitation of intimate thoughts that enable the long-term generation of individual personality and the concomitant emotional behaviours that flow.

However, the private traits that are intimately related to the home also make the smart home a significant target of commercial interest. The newly developing smart homes are sites of dense sensorisation, replete with multiple sensorised devices and infrastructures that enable substantial data collection by multiple commercial parties.[4] Behind the seamless interaction of the smart home there lies a commercial imperative to collect sensorised data to identify patterns and behaviours of

---

[1] Ruth Gavison, 'Privacy and the Limits of Law' (1980) 89 Yale Law Journal 421, 464. Lisa Austin, 'Re-reading Westin' (2019) 20 Theoretical Inquiries in Law 1.

[2] *Griswold v Connecticut*, 485 regarding the seminal Supreme Court decision on Fourth Amendment considerations in the home; Anita Allen, *Unpopular Privacy: What Must We Hide?* (Oxford University Press 2012) 17 regarding legal protections accorded to activities in the home that would be illegal in other spaces.

[3] Ken Gormley, 'One Hundred Years of Privacy' [1992] Wisconsin Law Review, 1384 citing Justice Brennan in *Carey v Brown* 447 US 455 (1980) 471.

[4] Nazmiye Balta-Ozkan and others, 'The Development of Smart Homes Market in the UK' (2013) 60 Energy 361.

individuals at their most natural and raw. Smart home collections provide data to garner prescriptive nudges based on the inference of future individual desires. Seamless digital operations and the collections of behavioural data are thus inseparable in the smart home.[5] To put it another way, if personal information is indeed the new oil,[6] then the smart home is the primary oil well in the fields of collected future.

Understanding the technical framework of the smart home is important as it reveals a complex data collection and communication environment that has multiple data pathways and numerous data collection components. This chapter outlines the development of the smart home, which has roots dug deep into consumerist soil. It also examines some of the challenges faced in smart home development that were largely bypassed in the 2000s by the convergence of Wi-Fi networks, new communication protocols and extended data storage capacities. The convergence has enabled the development of smart home possibilities driven by sensorised devices, home components and infrastructures. All of this starts to enable the seemingly seamless nature of the smart home to flourish.

However, behind the smart home's seamless fascia reside several deeply embedded data collection and sense-making infrastructures. These are brought to the surface by considering the smart home's sensing, reasoning and intervening processes. The sensing process requires an anatomy of the smart home's sensor data generation and collection capabilities. As argued throughout the book, the collected world is understood properly through the collection of sensor data. In the smart home, sensor data collections are complex, multiple and fragmented. Our discussion of the reasoning process examines the sense-making capacities of predictive and prescriptive forms of data analytics. In keeping with the book's focus on collection consequences, greater attention is paid to the use of collected sensor data than to the algorithmic models that attempt to make sense of that data. Finally, in looking at the intervening process, we consider the application of seamless prescriptive nudges in the smart home through the fast-developing area of smart home insurance. In this regard, the smart home's intervening process is also used to introduce some of the key elements of smart home business models, based on sensor data collection and covered in greater depth in Chapter 4.

---

5 Steven Weber and Richmond Y Wong, 'The New World of Data: Four Provocations on the Internet of Things' (2017) 22 First Monday 1, 5.
6 Editorial, 'Fuel of the Future' *The Economist* (London, 6 May 2017) Briefing <www.economist.com/news/briefing/21721634-how-it-shaping-up-data-giving-rise-new-economy> accessed 2 May 2019.

3.2   Key Smart Home Developments

We tend to think of the smart home as a modern, technologically focused development. However, the genesis of the smart home finds its roots in the growth of consumer devices for the home in the post-war period of the 1950s and 1960s.[7] These roots, in turn, find their initial flourishing in the 1920s with the increasing electrification of the home.[8] Device developments became increasingly intertwined with the vision of a home as a seamless consumer haven of technological activity.[9] Pre-internet, smart home commercial activity was oriented towards time-saving consumer goods, particularly aimed at the woman of the home.[10] Conceptualisations of the smart home began to move from the annals of science fiction to an increasing consumer reality in the 1980s. Product manufacturers increasingly experimented with the newly realised possibilities of two-way communications emerging from the protean Internet.[11] These possibilities give rise to new technological depictions of the video phone and other previously unimaginable devices.

Despite long-term interest in the notion of a seamless home, structural implications of smart home development only began to emanate seriously from the 1990s onwards – again, spurred on by the advent of the Internet.[12] Even then, however, the smart home did not manifest at scale due to significant structural barriers.[13] These include the cost of required infrastructural development,[14] particularly in technological substructures that are dependent upon physically upgrading older housing stock with new forms of cabling.[15] The lack of a common protocol for smart home development was also a significant roadblock[16] and is

---

[7] Richard Harper, 'From Smart Home to Connected Home' in Richard Harper (ed), *The Connected Home: The Future of Domestic Life* (Springer 2011).

[8] Aldrich 19.

[9] Michael Porter and James Heppelmann, 'Connected Products Are Transforming Competition' (2014) Harvard Business Review 1, 9.

[10] Lynne Baillie and David Benyon, 'Place and Technology in the Home' (2008) 17 Computer Supported Cooperative Work 227; Charlie Wilson, Tom Hargreaves and Richard Hauxwell-Baldwin, 'Smart Homes and Their Users: A Systematic Analysis and Key Challenges' (2015) 19 Personal and Ubiquitous Computing 463, 467.

[11] Aldrich 22.

[12] Ibid.

[13] Nazmiye Balta-Ozkan and others, 'Social Barriers to the Adoption of Smart Homes' (2013) 63 Energy Policy 363; Charlie Wilson, Tom Hargreaves and Richard Hauxwell-Baldwin, 'Benefits and Risks of Smart Home Technologies' (2017) 103 Energy Policy 72.

[14] J Brich and others, 'Exploring End User Programming Needs in Home Automation' (2017) 24 ACM Transactions on Computer-Human Interaction 1.

[15] Aldrich 22.

[16] Biljana L Risteska Stojkoska and Kire V Trivodaliev, 'A Review of Internet of Things for Smart Home: Challenges and Solutions' (2016) 140 Journal of Cleaner Production 1454.

still pertinent to today's considerations, as highlighted in the discussion that follows. Moreover, despite the early experimentations of consumer device manufacturers, there was little attempt to match technologically focused products with actual consumer needs.[17] In other words, there was a misalignment between manufacturing incentives and consumer markets for smart home products.[18] All of this made viable business model development for smart homes problematic.[19]

The situation changed dramatically at the start of this century, when the actuality of the smart home started to be realised. The resurgence of the smart home from the late 2000s was heavily linked with the convergence of several technological developments.[20] Increases in wireless bandwidth, particularly in home-based Wi-Fi networks, enabled the development of new, sensorised home devices.[21] In turn, the reduced cost of sensor manufacture provided a production platform that led to the increasing sensorisation of home-based devices and infrastructures. Finally, the reduced cost of new forms of sensorised data collection, acquisition and storage[22] finally enabled the prospect of the seamless home. Some of the previously intractable roadblocks were now removed. Large-scale Wi-Fi availability simply bypassed the previously significant cost inhibitor of cable rewiring, and new forms of cloud data warehousing vastly reduced the cost of data storage. Newly sensorised devices massively increased the possibilities of data generation and collection.

The foundational stage was now set for rapid development. The effect of smart home technological convergence averted the requirement for new specialised technical environments. Fledgling smart home business models moved towards a new data collection target: the behaviours and activities of adopted technology by participants in the home.[23] Corporations realised that the increasingly ubiquitous use of

---

[17] L Takayama and others, 'Making Technology Homey: Finding Sources of Satisfaction and Meaning in Home Automation' (UbiComp'12 – Proceedings of the 2012 ACM Conference on Ubiquitous Computing, Pittsburgh, September 2012).

[18] Aldrich 22.

[19] James Barlow and Tim Venables, 'Smart Home, Dumb Suppliers? The Future of Smart Homes Markets' in Richard Harper (ed), *Inside the Smart Home* (Springer 2003) 260.

[20] David Barnard-Wills, Louis Marinos and Silvia Portesi, *Threat Landscape and Good Practice Guide for Smart Home and Converged Media* (ENISA 2014) 7.

[21] Balta-Ozkan and others, 'The Development of Smart Homes Market in the UK' 363–4.

[22] Guilherme Mussi Toschi, Leonardo Barreto Campos and Carlos Eduardo Cugnasca, 'Home Automation Networks: A Survey' (2017) 50 Computer Standards & Interfaces 42.

[23] Simon CR Lewis, 'Energy in the Smart Home' in Richard Harper (ed), *The Connected Home: The Future of Domestic Life* (Springer 2011); Porter and Heppelmann 8.

home Wi-Fi provided new data collection possibilities.[24] A new market for smart home sensorised devices and infrastructures suddenly unfolded, giving rise to a number of new technological developments. A data and device symbiosis started to emerge that had previously been impossible. Manufacturers realised that they now had the capacity to collect data about how their products were being used, because they could now collect near-to-real-time data on product usage. The historical misalignment outlined previously was now replaced by a direct alignment between manufacturers, service providers and the users of their products.

The realignment fuelled an explosion of new sensorised products and services for the smart home, such as the number of home control systems that have recently entered the market. These include initial 'start-ups' such as pioneers Tado and Nest but also some existing technology and electronic home device giants such as Phillips, Samsung and LG. The major tech titans – particularly Amazon, Google and Facebook – also see the commercialisation possibilities from smart home–focused voice control systems, as highlighted later. All of this augurs well for the upcoming commercial battle for supremacy of the smart home, and its control systems, as evidenced by the smart thermostat.

Tado is an Internet-controlled thermostat that automatically adjusts home heating based on whether there is a registered human presence in the home and other factors such as weather.[25] Human presence is assessed by the location of a user's smartphone. If the phone's location is confirmed as being in the house, then the Tado system adjusts the heating and air conditioning. Alternatively, if no registered phone is in the home, then the air conditioning is switched off and it is switched back on when the phone returns within a specified distance of the home. To function, Tado requires a smartphone app connected to a wireless device, which then connects to the air conditioning unit via infrared signal. Thus, the smartphone replaces the old remote as the primary control device. Nest is a similar technology and provides smart homes with 'learning thermostats' that can assess when a home is empty and how long it takes for a house to warm through.[26] A security webcam can also be used for the purpose of visually monitoring a home if the owner is absent and to provide custom alerts based on certain movements around

[24] Balta-Ozkan and others, 'Social Barriers to the Adoption of Smart Homes' 368.
[25] 'Home' (Tado 2019) <www.tado.com/us/> accessed 20 June 2019.
[26] 'Thermostat' (Nest 2019) <https://store.nest.com/product/thermostat/T3007ES> accessed 20 June 2019.

doors or windows.[27] The Nest system is covered in depth in Chapter 4 as part of the coverage of smart home insurance data collection models.

House lighting is another overlooked home feature that can lead to unnecessary expense without knowledgeable regulation. Standard light fittings are operated by a flick of the switch and a light goes on or off, with minimal possibility of adjustment or control. Mostly, it is a binary action. The switch goes on or off and the light goes on or off. Lighting in the smart home is fundamentally different. The Phillips Hue range of LED light bulbs provides an illuminating example.[28] A smartphone can now be used to switch the light on and off. Moreover, the lights themselves can be changed to many different hues of white, or even different colours, and can flash, pulse or dim. The smartphone app connects to the light bulbs via a wireless router, and it is possible to activate lights without being present in the home. Up to fifty light bulbs can be triggered at any given time, either collectively or individually.

Given the increasing smartness of different home components, the next target for tech giants has been one-touch control of all smart home devices. In 2014, Apple announced the release of HomeKit, a new network protocol that allowed home users to control all of their smart home devices through an iPhone or iPad.[29] A HomeKit user can create a number of 'scenes' that can lead to the automatic activation of certain home devices or fittings. For example, a 'waking up' scene can be programmed to ensure that room temperature, lighting and curtain opening all work in sync.[30] A user can also program 'custom triggers' that alter scene activities based on a range of factors such as time and user location or user activity.[31] In relation to the 'waking up' scene, if the user sleeps past a certain specified time, then the custom trigger will increase the speed at which the lights are fully engaged and the curtains are completely opened. A smart, self-directed punishment for sleeping in can therefore be programmed.

---

[27] 'Nest Cam Indoor' (Nest 2019) <https://store.nest.com/product/camera/NC1102ES> accessed 20 June 2019.

[28] 'Light Your Home Smarter with Philips Hue' (Phillips 2019) <www2.meethue.com/en-au/philips-hue-benefits> accessed 20 June 2019.

[29] 'Your Home at Your Command' (Apple 2019) <www.apple.com/ios/home/> accessed 20 June 2019.

[30] Christina Warren, 'With iOS 9, Homekit Will Get Smarter and More Powerful' (*Mashable Australia*, 2015) <http://mashable.com/2015/06/10/ios-9-homekit-wwdc/#HbBcFkW23qqZ> accessed 20 June 2019.

[31] Darrell Etherington, 'HomeKit in iOS 9 Supports Custom Triggers, Apple Watch without iPhone and More' (*Tech Crunch*, 2015) <http://techcrunch.com/2015/06/10/homekit-in-ios-9-supports-custom-triggers-apple-watch-without-iphone-and-more/> accessed 20 June 2019.

Home control systems such as HomeKit become increasingly important given the rapid growth of smart devices for the home. As with the smartphone, electronic devices for the home are smarter because of their sensorisation. Those sensors provide greater opportunities for more user control and service delivery. For instance, the smart television can now listen to user verbal commands or accept the input of physical gestures and respond thanks to the television's sensors, the microphone and the camera.[32] Home control systems allow the smart television to be activated upon certain conditions, such as when a person returns home or at a certain time of day or night.

The smart fridge can also be controlled by smartphone. LG's fridge app, Smart Manager, can be used to check what is inside a fridge without opening the fridge door.[33] Moreover, the LG fridge will have Amazon Alexa installed and can be used for online grocery purchases; some items can be purchased automatically when the fridge calculates that supplies are running low. The next generation of LG smart washing machines will be able send an alert message when detergent is running low and recommend wash cycles based on the clothes being cleaned.[34] The smart vacuum cleaner remembers what it has cleaned and exhibits its cleaning history for the homeowner to check, in case they have forgotten their own vacuuming activities. Peggy, the smart clothes peg, can inform its owner that the washing on the line is dry or send a warning that it is about to rain.[35]

It should be noted that these device developments have unfolded in a remarkably short space of time, over the past decade. The past five years have also seen the advent of a device which is likely to have a long-term impact on smart home developments – the smart speaker, which has increasingly become a more generalised home voice assistant. Voice assistants were initially designed to enable voice-activated music playing and thus had limited scope in the context of smart home development. However, the significance of voice control assistants in the smart home is dramatically changing, because the assistant can now be used as a de facto home automation control system.

The first device of this type to arrive on the market was Amazon Echo, which featured Alexa as its virtual assistant. Amazon Echo first

---

[32] Damon Beres, 'How to Stop Your Smart TV from Eavesdropping on You' (*The Huffington Post*, 2015) <www.huffingtonpost.com.au/entry/your-samsung-tv-is-spying-on-you_n_6647762.html?section=australia> accessed 20 June 2019.

[33] 'Cool Meets Smart' (LG Smart ThinQ, 2019) <www.lg.com/us/discover/smart-thinq/refrigerators> accessed 20 June 2019.

[34] 'LG TWIN Wash' (LG Smart ThinQ, 2019) <www.lg.com/us/lg-thinq-appliances/products/lg-twinwash-thinq/index.html> accessed 20 June 2019.

[35] 'Meet Peggy, the World's Smartest Clothes Peg' (OMO 2019) <www.omo.com/au/dirt-is-good/real-play/peggy.html> accessed 20 June 2019.

launched in 2014 in the United States and has since spread world-wide. By the beginning of 2019, estimates highlighted that Amazon had sold over 100 million Alexa-powered devices in its Echo product line,[36] including Echo Dot, which is Amazon's version of a mini smart speaker and has the same capabilities of Echo. In addition, the Echo Show has a video touchscreen. Google Home is the other major voice control assistant pertinent to smart home development. It was launched in late 2016 in the United States, and like the Amazon Echo, it initially appeared as a smart speaker that has since developed and spread across the globe. Google Home is powered by its voice recognition software, Google Assistant, and operates similarly to the Amazon Echo. By mid-2018, it was estimated that Google had sold 52 million Home devices[37] throughout the world, with the greatest uptake in the United States.[38]

The other significant development to the Google Home family is Google Home Hub, released in 2018, which is similar in design and purpose to the Amazon Show.[39] Hub, like Show, brings together the benefits of a voice controller with a basic tablet that can be used as a voice assistant, a digital photo frame or a smart home control panel.[40] The latter use is particularly important because it seems to indicate Amazon's and Google's interest in being significant players in the smart home control market, which has important implications for business models based on smart home collections, as highlighted in Chapter 4.

Amazon and Google appear to dominate market share for home voice assistants.[41] However, other major technology companies are also seeking to enter the market. Apple's HomePod is powered by the Siri voice assistant and is capable of voice controlling other home devices

[36] Abrar Al-Heeti, 'Amazon Has Sold More Than 100 Million Alexa Devices' (*CNet*, 2019) <www.cnet.com/news/amazon-has-sold-more-than-100-million-alexa-devices/> accessed 20 June 2019.

[37] Google also has a smaller smart speaker version of Google Home, Google Home Mini.

[38] Jillian D'Onfro, 'Google's Small Hardware Business Is Shaping Up, Could Book $20 Billion in Sales by 2021, RBC Says' (*CNBC*, 2018) <www.cnbc.com/2018/12/21/google-hardware-revenue-profit-potential-rbc-analyst-mark-mahaney.html> accessed 20 June 2019.

[39] 'Introducing Nest Hub' (Google Nest 2019) <https://store.google.com/product/google_home_hub> accessed 20 June 2019.

[40] Andrew Gebhart, 'Google Home Hub Review: Google Assistant Helps This Tiny Screen Feel Powerful' (*CNet*, 2018) <www.cnet.com/reviews/google-home-hub-review/> accessed 20 June 2019.

[41] Will Oremus, 'Alexa Is Losing Her Edge' (*Slate*, 2018) <https://slate.com/technology/2018/08/amazon-echo-is-losing-smart-speaker-market-share-to-google-home-heres-why.html> accessed 20 June 2019.

and components through Apple HomeKit.[42] In 2018, Samsung entered the smart speaker market with the release of Galaxy Home, which acts primarily as a speaker that can also be used to voice control Samsung devices on its SmartThings platform.[43] Finally, in early 2019, Facebook confirmed that it is working on a voice control assistant to rival Alexa, Google Assistant and Siri as part of a new range of product development,[44] including Portal, a device primarily designed for video calls using Facebook Messenger and for playing music.[45]

These developments highlight that smart home ideation is heavily influenced by the notion of seamless user control of the home through a range of different sensorised devices. However, it would be wrong to think of the smart home as a clearly defined construct. One group of researchers has called the burgeoning literature on smart homes 'utterly incoherent'.[46] Not surprisingly, various definitions of a smart, connected or digital home exist and are pertinent to three broad perspectives of a smart home: functionality, instrumentality and socio-technicality.[47]

The key difference between these perspectives regards differing norms of control.[48] Under the functionality perspective, 'control' refers to enhanced consumer uses of the home through various devices. Seamlessness according to this perspective is dominant and inherently tied to notions of consumerism. Control under the instrumentality perspective regards government macro control and how the smart home can be utilised to meet a broader policy agenda, such as energy

[42] Julie Clover, 'How to Use HomePod to Control Your HomeKit Devices' (*MacRumors*, 2018) <www.macrumors.com/how-to/homepod-control-homekit-devices/> accessed 20 June 2019.

[43] Jonathan Chadwick, 'Galaxy Home: Can Samsung Make Up for Lost Time in the Smart Speaker Space?' (*ZDNet*, 2018) <www.zdnet.com/article/galaxy-home-can-samsung-make-up-for-lost-time-in-the-smart-speaker-space/> accessed 20 June 2019.

[44] Nick Statt, 'Facebook Confirms It's Working on an AI Voice Assistant for Portal and Oculus Products' (*The Verge*, 2018) <www.theverge.com/2019/4/17/18412757/facebook-ai-voice-assistant-portal-oculus-vr-ar-products> accessed 20 June 2019.

[45] Dan Seifert, 'Facebook Portal Review: Trust Fail' (*The Verge*, 2018) <www.theverge.com/2018/11/8/18072998/facebook-portal-plus-smart-display-messenger-review-price-specs> accessed 20 June 2019.

[46] Sam Solaimani, Wally Keijzer-Broers and Harry Bouwman, 'What We Do – and Don't – Know about the Smart Home: An Analysis of the Smart Home Literature' (2015) 24 Indoor and Built Environment 370, 371.

[47] Wilson, Hargreaves and Hauxwell-Baldwin, 'Smart Homes and Their Users'.

[48] Scott Davidoff and others, 'Principles of Smart Home Control' in Paul Dourish and Adrian Friday (eds), *UbiComp 2006: Ubiquitous Computing* (Springer Berlin Heidelberg 2006).

efficiency[49] or home-based healthcare,[50] particularly for aged health-care monitoring.[51] Finally, the socio-technical perspective regards the changing role of decision making in the home and the consequences of transferring decisions from humans to machines based on policy considerations and individual behaviours.

The large range of smart home definitions also indicate conceptual incoherence about the smart home's role and purpose, which cut across all three perspectives. Definitions encapsulate different aspects of the smart home, ranging from its ability to provide seamless control to a deeper consideration of its technological infrastructures. The smart home enhances monitoring and control functionality into homes,[52] and this control functionality supports inhabitants in their daily lives via technological means from inside and outside the home.[53] Enhanced control afforded through automated technology means that the smart home can anticipate and respond to the needs of occupants in order to promote convenience[54] and quality of life enabled through the facilitation of more accurate monitoring.[55] The interconnection of various sensorised technologies[56] is a key constituent of smart home automation that covers a range of different devices and can be controlled in many different ways.[57]

The notion of control through devices and the automated consequences is best captured by Francis Aldrich and his influential categorisation framework that adduces five classes of smart home[58]:

[49] Ameena Saad Al-Sumaiti, Mohammed Hassan Ahmed and Magdy MA Salama, 'Smart Home Activities: A Literature Review' (2014) 42 Electric Power Components and Systems 294, 295.

[50] Brent Mittelstadt, 'Designing the Health-Related Internet of Things: Ethical Principles and Guidelines' (2017) 8 Information 77.

[51] G Demiris and BK Hensel, 'Smart Homes for Patients at the End of Life' (2009) 23 Journal of Housing For the Elderly 106.

[52] Wilson, Hargreaves and Hauxwell-Baldwin, 'Benefits and Risks of Smart Home Technologies' 73.

[53] MA Latif and others, 'User Privacy Framework for Web-of-Objects Based Smart Home Services' (2015) 9 International Journal of Smart Home 61.

[54] Balta-Ozkan and others, 'The Development of Smart Homes Market in the UK' 369.

[55] G Demiris and BK Hensel, 'Technologies for an Aging Society: A Systematic Review of "Smart Home" Applications' [2008] IMIA Yearbook of Medical Informatics 33.

[56] D Geneiatakis and others, 'Security and Privacy Issues for an IoT Based Smart Home' (40th International Convention on Information and Communication Technology, Electronics and Microelectronics (MIPRO), Opatijia, Croatia, May 2017); Joseph Bugeja, A Jacobsson and P Davidsson, 'On Privacy and Security Challenges in Smart Connected Homes' (European Intelligence and Security Informatics Conference, Upsalla, Sweden, August 2016).

[57] Toschi, Campos and Cugnasca 45.

[58] Aldrich 35.

- *Homes which contain intelligent objects.* These homes contain single, standalone appliances and objects which can function in an intelligent manner.
- *Homes which contain intelligent, communicating objects.* These homes contain appliances and objects which function intelligently in their own way and exchange information between one another to increase the overall functionality of the home.
- *Connected homes.* These homes have internal and external networks, allowing interactive and remote control of systems, as well as access to external services and information through the Internet.
- *Learning homes.* These homes can record and accumulate data about patterns of activity that are then used to anticipate user needs and to inform about activities and device usages.
- *Attentive homes.* These homes can monitor the activity and location of people and objects within homes and this information is used to automate control technology in anticipation of the occupants' needs.

Aldrich's framework distinguishes between a home automation system that can learn and one that cannot, and one that is able to maintain constant awareness of participants and one that is not. The framework is also hierarchical from a technical perspective, as each class generally has systems in place from the previous class. Aldrich's analysis is helpful because it highlights the interconnected nature of the smart home, which manifests in vast forms of data collection. This point is exemplified and helpfully extended by Simon Lewis in his conceptualisation of the smart home as a 'digital nervous system' that is

> capable of *sensing* the state of many variables [in the home], *reasoning* about the conditions that prevail, and then *intervening* to deliver some desired improvement or optimization.[59]

Lewis' definition captures the core conceptual facets of the smart home explored in this book and details the different data collection and analysis states required for the home to become smart. The sensing state refers to the sensorised data collections through the increased use of sensorised devices and their interconnection. The reasoning state refers to the analysis of collected smart home data that can enable the inference and detection of predicted user behaviours. Finally, the intervening state provides the prescriptive output for automated optimisation based on the prediction of individual behaviours and the utilisation of home resources.

[59] Simon CR Lewis, 'Energy in the Smart Home' in Richard Harper (ed), *The Connected Home: The Future of Domestic Life* (Springer 2011).

These three states are now examined further, beginning first by identifying the complex and fragmented infrastructural anatomy of the smart home.

3.3   Sensing: The Infrastructural Anatomy

For the smart home to sense, reason and intervene, several interconnected components are required to facilitate data collection, exchange and analysis. The resurgence in smart home business model development has increased the availability of smart home products and systems, as outlined previously. However, the current state of the smart home market is still relatively immature. A key factor in the low level of maturity is the complexity of smart home infrastructures. The removal of historical roadblocks has opened market opportunities, but at present there is no single dominant technological framework or smart home model for commercialising sensorised data.[60] Instead, a vast array of devices, manufacturers, protocols and data collection pathways exist. It should be no surprise, therefore, that there is not an agreed-upon vocabulary that describes the smart home environment.[61] The smart home market is developing rapidly, but it may well be the case that it continues to be multifaceted and fragmented.[62] That said, the following components appear to be important constituents of technological frameworks of home control/automation systems. Figure 3.1 provides a diagram of the infrastructural components of the smart home. The diagram, and the analysis that flows, details the data collection frameworks of home automation systems.

3.3.1   Controllers

Section 3.3 highlighted the intimate link between the smart home and functional control of its components. The ability to control smart home devices and infrastructures resides in a plethora of different controllers. 'Controllers', in the smart home sense, refers to different control devices that a smart home participant can use to operate sensorised home products or home infrastructures, such as air conditioning, security and utilities.[63] The number of user control options in the smart

---

[60] Wilson, Hargreaves and Hauxwell-Baldwin, 'Benefits and Risks of Smart Home Technologies' 80.
[61] Saad Al-Sumaiti, Ahmed and Salama.
[62] Balta-Ozkan and others, 'The Development of Smart Homes Market in the UK'.
[63] Saad Al-Sumaiti, Ahmed and Salama 301.

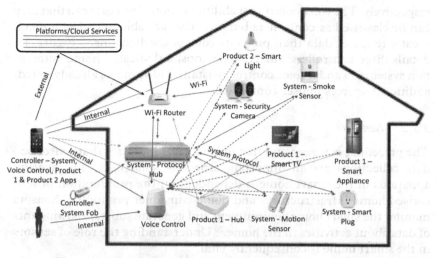

Figure 3.1 Smart home component architecture and data collection.

home can be significant and is largely dependent upon individual product options and the home automation/control system installed.[64]

Different products may each provide their own controller, such as a key fob, a remote controller or a smartphone app. It is also possible that the same product can be operated by a number of different controllers.[65] For example, Samsung smart home products can be controlled by a dedicated SmartThings hub, a separate remote controller for a specific product, a smartphone app or via Galaxy Home. To add to the complexity of control options, smart home participants also interact with smart home devices in different ways, and often in combination with different controllers.[66] The seemingly innocuous controller plays an important role in smart home data collections, as they are the primary point of human agency in the control of the home. In terms of data acquisition and analysis, data generated from participant controller use appears to be important because it provides an insight into user home control activities and behaviours.[67] This factor may also help to explain Google's and Amazon's significant backing of Home and Echo,

[64] Eric Zeng, Shrirang Mare and Franziska Roesner, 'End User Security & Privacy Concerns with Smart Homes' (Symposium on Usable and Scable Privacy (SOUPS), Santa Clara, California).
[65] Geneiatakis and others 1293.
[66] Zeng, Mare and Roesner.
[67] Andreas Jacobsson, Martin Boldt and Bengt Carlsson, 'A Risk Analysis of a Smart Home Automation System' (2016) 56 Future Generation Computer Systems 719.

respectively. The voice control capabilities of both devices mean that they can be classified as controllers but ones that are able to collect a much greater range of data than previous control mechanisms.[68] Figure 3.1 details three controllers: smartphone apps, a dedicated home automation system fob and a voice control assistant. However, as already noted, additional sources of user control exist.

3.3.2   Sensors

The preceding coverage and that of Chapter 2 indicates that sensorised data collections are an important aspect of smartness. The sensing apparatus of the smart home is dependent on the proliferation of sensorised home infrastructures and home consumer products.[69] Sensors monitor the smart home environment and generate substantial amounts of data about activities in the home.[70] Understanding the role of sensors in the smart home is consequently vital.

The type of sensors employed can have a significant impact upon the type of home control or automation system in place. Smaller sensors – typically those used for movement monitoring of doors and windows – have limited battery and data exchange capabilities.[71] Data collection and exchange capacities are minimal and thus require specific smart home protocols to function. Larger sensors, on the other hand, such as video cameras, have much greater battery and data exchange capabilities, and data exchange is undertaken via internal Wi-Fi networks or the Internet.[72] Products with smaller sensors typically have specific smart home protocol controllers,[73] whereas infrastructures or products with larger sensors can be operated via smartphone app or other mobile devices.

These different sensors become important because they require different communication pathways and infrastructures needed to enable sensorised collections. The data collection capacities of a smart home will therefore be dependent upon the installation and use of sensors in that home.

---

[68]   Vijay Sivaraman, Habibi Hassan Gharakheili and Clinton Fernandes, *Inside Job: Security and Privacy Threats for Smart-Home IoT Devices* (2017).
[69]   Harper, 'From Smart Home to Connected Home' 13.
[70]   Andreas Jacobsson and Paul Davidsson, 'Towards a Model of Privacy and Security for Smart Homes' (IEEE 2nd World Forum on Internet of Things (WF-IoT), Milan, December 2015).
[71]   Barnard-Wills, Marinos and Portesi 19.
[72]   Ibid.
[73]   Geneiatakis and others 1293.

3.3.3   Smart Home Communication Protocols

As highlighted previously, certain sensors require specific smart home communication protocols. These protocols provide a designated language for data exchange involving low-battery-life sensors.[74] Most smart homes now have Wi-Fi; however, Wi-Fi data exchange is a considerable burden on sensor batteries which is felt across the whole smart home system. The more sensorised devices that connect to a home's Wi-Fi system, the slower the system will become. Specific smart home low-energy data transfer protocols have thus developed to ameliorate the problems with Wi-Fi- and Bluetooth-based sensorised data collections.[75]

There are several different protocols in existence – such as Z-Wave, Zigbee, Insteon and HomeKit Accessory Protocol.[76] To fragment the situation further, specific smart home control or automation systems generally operate on specified protocols. For example, as detailed in Chapter 4, the smart home automation system Fibraro operates solely with Z-Wave; the Amazon Echo operates with Wi-Fi and Zigbee; and Nest Protect operates solely with its own proprietary protocol, Weave. To complicate the situation further, depending on the sensorised product, it may also be possible to operate a smart home protocol-based system via smartphone or Wi-Fi, depending on the type of hub in operation.[77]

Let's recap before we move on. Smart homes can have multiple controllers. These multiple controllers can control individually or collectively a different range of sensorised devices and infrastructures. Smart home devices have different types of sensors with different battery and data exchange capabilities. Certain sensors require dedicated smart home communication protocols to collect data. Different devices and smart home systems and controls operate on different protocols. These different protocols can be dedicated to one type of product, or they can be used by different controllers. It thus becomes clear that there is not a 'one size fits all' smart home model of data generation or collection. In fact, the opposite is the case.

---

[74] Bugeja, Jacobsson and Davidsson 173.
[75] Ibid.
[76] Huichen Lin and Neil Bergmann, 'IoT Privacy and Security Challenges for Smart Home Environments' (2016) 7 Information 44.
[77] Geneiatakis and others 1293.

3.3.4   *Protocol Hub*

The protocol hub is the brain of the smart home's digital nervous system. The hub can also be called the 'control system' or 'gateway'. It is the smart home's central repository of data communication, acquisition and exchange.[78] The hub serves several purposes: it collects sensor-generated data, it facilitates the interoperability of different products using the same protocol, it links protocol-based products/infrastructures with smartphone capabilities and it provides a communications gateway to the outside world via Wi-Fi.[79]

Several different home control/automation system providers come with their own hubs – for example, Fibraro, VeraEdge, Aeotec and Wink – as do product manufacturers such as Samsung, Philips Hue Bridge and Internet providers (for example, Telstra in Australia). Depending on the protocol in operation, some products may also be connectable to some other home control or automation systems. For example, Samsung products can connect to a Wink hub. However, some other products may not connect to the same hub because they use different smart home protocols.[80]

To complicate further the data communication pathways, most smart home protocols cannot connect directly to the Internet, and thus the hub provides a communication link to Wi-Fi or the Internet that enables smartphone control and use of protocol-based products. The Samsung SmartThings hub is an example of a smart home hub that has external communication capabilities. However, adding to the complexity is the fact that some home control or automation systems and devices have been designed to communicate directly with Wi-Fi and thus obviate the need for a protocol hub.[81] These include, for example, Belkin Wemo, D-Link Connected Home and Nest.

The key connectivity factor, as discussed earlier, regards the type of sensor in operation and its requisite battery life.[82] Figure 3.1 details the different connection system pathways where battery-powered sensorised devices connect directly to the hub (e.g. motion sensor) whereas larger sensors connect to the hub via Wi-Fi (e.g. security camera). The existence or lack of a protocol hub is an important factor in data generation considerations for smart home data collection business models, as detailed in Chapter 4.

---

[78] Risteska Stojkoska and Trivodaliev 1458.
[79] Tiago DP Mendes and others, 'Smart Home Communication Technologies and Applications: Wireless Protocol Assessment for Home Area Network Resources' (2015) 8 Energies 7279.
[80] Geneiatakis and others.
[81] Ibid.
[82] Risteska Stojkoska and Trivodaliev.

3.3.5   Wi-Fi Router

Given the foregoing discussion and the multifaceted forms of connectivity applicable, the home Wi-Fi router is also an important component in home control or automation systems. The router can provide direct connection to the Internet, which facilitates smart device user control and external sensorised data transfer and exchange. Figure 3.1, for example, details the communication pathway where a customer controls the home device outside of the home. The router can also be the communications link for the protocol hub to enable external data transfer, such as in the security camera example in Figure 3.1.

3.3.6   Smart Home Platforms

The final component to note is the role of smart home platforms or cloud services. Again, to encapsulate the complexity of the smart home environment, these platforms are still in formulation, and there is no single dominant platform or even one type of dominant platform. A smart home platform appears to serve three primary functions – namely to

1. provide a protocol format for applications including data generation, storage and analysis of operations;
2. match applications and services with users based on analysis of the data generated; and
3. automate and adjust smart home products or infrastructures based on user behaviours.

The predominance of platform function appears to shape different data generation business models and relates to the connective basis of the smart home. Predominant function 1 platforms tend to be protocol based – for example, the Z-Wave Alliance – whereas predominant functions 2 and 3 are not specifically protocol based and thus are not dependant on a protocol hub for data generation activities.

It should also be noted that some platforms can be defined as open or versatile, such as Samsung SmartThings, which is compatible with Wi-Fi-, Bluetooth-, Z-Wave- and Zigbee-based devices. Others are closed and controlled, such as Tesltra's iControl Platform, which only operates on the Zigbee protocol and with any other devices that Telstra has specifically partnered with. The emerging voice control home systems – particularly Amazon Echo and Google Home – will also significantly impact the development of smart home platform-based business models in the future given the apparent favoured use of voice assistants by smart home consumers.[83] Figure 3.1 highlights the greater possibilities for device connection offered by the platform home voice control systems.

3.4   Reasoning: The Analytical Processes

In this chapter, we have thus far looked at the resurgent smart home, restimulated by technological developments, and outlined the smart home's infrastructural anatomy to highlight the complex data generation infrastructures that pervade beneath the surface. Behind this invisible technological structure is the desire to generate and collect data about the home and the activities within it. Data generation in the home is giving rise to a range of new business logics that are increasingly focused on these invisible data collection structures of sensorisation to capture accurate recordings of routines and patterns.[84] The issue consequently shifts from the sense-generating activities of the smart home to the sense-making state of reasoning in Lewis' conceptual framework. The second state heralds the advent of predictive analytics and the increasing onset of prescriptive analytics for making sense of smart home sensorised data. A brief review of data analytic considerations as an integral process of smart home reasoning is required.

The sophistication of any data analytical task is dependent upon the sophistication of data collection processes, analytical processes and the development of metrics and models.[85] The adage of garbage in, garbage out is very much pertinent in the collected environment of the smart home. The data generation and collection processes of the smart home were outlined significantly earlier on. As demonstrated, the smart home is a complex data generation environment that is fragmented across several different axes. It is also important to consider the implications of metadata generation, as significant insight can be derived from metadata analysis.

Metadata is, of course, data about data; in the context of the smart home, 'metadata generation' refers to automated data produced by the sensorised devices and smart home systems about their use. These data 'breadcrumbs' provide a treasure trove of information, because they essentially compose a log of activity in the smart home that can be used to identify and infer behavioural patterns.[86] For example, as detailed

---

[83] Zeng, Mare and Roesner.
[84] Wilson, Hargreaves and Hauxwell-Baldwin, 'Benefits and Risks of Smart Home Technologies'.
[85] Thomas H Davenport, Paul Barth and Randy Bean, 'How Big Data Is Different' (2012) 54 MIT Sloan Management Review 43.
[86] Jacobsson, Boldt and Carlsson 721.

later in this section, risk assessment inferences are now being made from metadata about battery-life activities in the home. In other words, ignoring low-battery warnings from devices such as sensorised smoke alarms will increasingly be modelled as risky behaviour.

Before data can be used for analytical purposes, however, it will generally have to be 'cleansed'. Data cleansing is to a certain extent the forgotten process of data analytics, and its purpose is often overlooked in favour of more exciting analytical processes – particularly predictive analytics. Data cleansing in this regard is nevertheless essential to the overall collection and analytical process, because it identifies incompatibility issues within datasets, such as incorrect data formats that could impact upon the validity of results. Furthermore, data cleansing may also involve anonymising personally identifiable data and creating unique identifiers to enable multiple dataset aggregation.[87] Again, in the smart home context, this could be the use of de-identified and aggregated data across a number of smart home device or system users.

As noted previously, different sensors have different data collection capacities. It is also important to note that, in general, two types of data are collected for analytical purposes in smart homes: structured and unstructured. Structured data such as data collected by sensors and the metadata derived produce largely similar text or numerical formats that make it easier to use for analytical purposes. Structured data tends to require less cleansing, as analytical processes will largely be designed around the availability of structured data from sensorised collections. De-identification subprocesses may be significant, though.[88]

Unstructured data, on the other hand, can come in many different formats. For example, unstructured data can include text, audio recordings, photographs and videos.[89] A common form of unstructured data in the smart home is video recordings from security cameras. Given the potential privacy sensitivities, smart home video is generally accorded higher levels of protection, as exemplified in the privacy policies of smart home systems detailed in Chapter 4. However, it is the combination of structured and unstructured data that provides the background

---

[87] Rahul Narain Saxena and Anand Srinivasan, *Business Analytics: A Practitioner's Guide* (Springer 2013) 38.

[88] Irv Lustig and others, 'The Analytics Journey: An IBM View of the Structured Data Analysis Landscape: Descriptive, Predictive and Prescriptive Analytics' (Analytics: Institute for Operations Research and the Management Sciences 2010) <www.analytics-magazine.org/november-december-2010/54-the-analytics-journey.html> accessed 20 June 2019.

[89] Intel IT Center, 'Big Data 101: Unstructured Data Analytics' (Intel 2012) <www.intel.com/content/dam/www/public/us/en/documents/solution-briefs/big-data-101-brief.pdf> accessed 20 June 2019.

for maximising predictive capabilities, because the combination of different data gives rise to unintuitive insights through unexpected correlations.[90] Meaning is given to these correlations in the form of predicted patterns generated by the final process, predictive analytics.

Predictive analytics entails the use of algorithms to data mine cleansed smart home system datasets in the search for unintuitive patterns. The search for unintuitive patterns is the goal of predictive analytics, so that new correlative insights can be gained from existing datasets that provide a new understanding of how a home is functioning and how it could function more effectively in the future. Predictive analytics transits descriptions of data, albeit fine-grained and detailed, to predictions of outcome.[91] In the smart home setting, this is the learning and attentive types of homes envisaged by Aldrich in which the smart home system can accurately predict the behaviours and needs of occupants and adjust home settings accordingly.

One of the key processes behind predictive analytics is segmentation. It involves the process of grouping together entities, or in this case customers of smart home devices, based on shared similarities, and it allows an organisation to identify and differentiate between different types of segments.[92] Segmentation allows organisations to learn more about the behaviour of certain groups within their overall customer cohort.[93] Once customer behaviours are more clearly identified, the organisation can then tailor and target resources, design choices and advertising strategies towards the behaviours of that specified segment.[94] Segmentation strategies become evident in the smart home context in Chapter 4 through the examination of smart home insurance models and the clear segmenting of customers based on certain categories of sensor data.

Predictive segmentation is the next step beyond traditional forms of segmentation that focus on descriptive reporting of existing data. The former type entails the situation where a customer is assigned to a segment based on the results of a predicted outcome.[95] For instance, rather than simply reporting whether a customer is a late payer, a predictive segmentation sequence could predict whether the late payer would be a

[90] Lustig and others.
[91] Saxena and Srinivasan 5.
[92] James Wu and Stephen Coggeshall, *Foundations of Predictive Analytics* (CRC Press 2012) 137.
[93] Thomas H Davenport, Jeanne Harris and Robert Morision, *Analytics at Work* (Harvard Business Press 2010) 83.
[94] Ibid 87 and the targeting elements of predictive analytics which result in the question 'Do we have a good target?'
[95] Ibid 83.

suitable customer to receive certain bonuses for customer loyalty based on a range of different correlated factors. A prior process of predictive analytics is consequently used to segment cohorts based on predicted outcomes. Predictive segmentation also creates another type of segmented scenario where a segment is identified but the identity or meaning of the group is yet to be established. In these situations, a predicted outcome is posited, and meaning is subsequently attached.

The next generation of data analytics, Analytics 3.0,[96] further extends the predictive elements of existing analytical frameworks through the advent of two significant developments: embedded processes and prescription.[97] In the 1950s, the Analytics 1.0 company focused on business intelligence and the use of information systems to aid understanding of company operations.[98] Data collection was cumbersome; decision making, as a consequence, was painstakingly slow and was further limited in scope by the restricted application of descriptive analytics and its inability to provide behavioural predictions.[99] By the mid-2000s, the Analytics 2.0 icons Google, Amazon and eBay radically extended the foundation of business intelligence through the use of 'business analytics'.[100] Unlike the Analytics 1.0 company that focused on core business data, all data is relevant for the Analytics 2.0 company, including new sources of data beyond internal company operations.[101] The vast new collections of data required new forms of data collection, and this in turn spurred technological developments that created new devices and processes which could better generate, collect, monitor and analyse data through predictive processes.

Analytics 3.0 now addresses a further phenomenon – the embedded employment of analytical processes in businesses and across industries. The Analytics 3.0 company attempts to record, collect and analyse everything about itself and its environment. Every internal organisational facet is measured and examined in the search for new unintuitive correlations that provide new insight into organisational functions.[102] In essence, the embedded nature of these developments also leads to

[96] Thomas H Davenport, 'Analytics 3.0' (2013) 91 Harvard Business Review 64.
[97] Ibid 69.
[98] HP Luhn, 'A Business Intelligence System' (1958) 2 IBM Journal of Research and Development 314.
[99] Bart Schermer, 'The Limits of Privacy in Automated Profiling and Data Mining' (2011) 27 Computer Law & Security Review 45.
[100] Thomas H Davenport, The New World of Business Analytics (International Institute of Analytics 2010); Lawrence Maisel and Gary Cokins, Predictive Business Analytics: Forward-Looking Capabilities to Improve Business Performance (John Wiley & Sons 2014).
[101] Davenport, 'Analytics 3.0' 67.
[102] Davenport, Harris and Morision.

new forms of logic. The Analytics 3.0 company must always, to be competitive, collect new streams of data to produce new correlated patterns that provide new insight.[103] In that sense, the Analytics 3.0 company is different from its predecessors, because it is infrastructural in nature both in terms of its requirement for industry-wide application and its embedded nature in organisational information infrastructures. The important point to consider here, addressed in Part III, is the situation in which Analytics 3.0 infrastructures effectively extend into the smart home through sensorised data collections as an intrinsic element of business model development, thus eradicating the misalignment of incentives between manufacturer and consumer highlighted earlier.

Analytics 3.0 also requires prescriptive analytics, which 'uses models to specify optimal behaviours and actions'.[104] Unlike predictive analytics, prescriptive analytics provides modelled solutions for predicted outcomes.[105] Prescriptive analytics is consequently the next step forward and regards the actual implementation of predictive projections into prescriptive outcomes that seek to modify human behaviours that maximise operational benefits.

Prescriptive analytics not only anticipates what will happen and when it will happen, but also why it will happen.[106] The focus of prescriptive analytics is therefore not solely the probabilistic prospects of predictive analytics. Instead, the focus is about generating possible solutions to facilitate the results of predicted outcomes. Prescriptive analytics is founded on optimisation which seeks to achieve the best outcome in acceptance with the complexity and uncertainty of existing information environments.[107] Thus, in the smart home context, the prescriptive outcome focuses on the development of the best, data-driven responses that maximise the use of home resources.

Prescriptive analytics is effectively the operationalisation, and potentially the automation, of decision making predicated on predictive outcomes and probabilised responses.[108] Moreover, the prescriptive models themselves become the means for embedding analytics into key

[103] Ibid.
[104] Thomas H Davenport and DJ Patil, 'Data Scientist: The Sexiest Job of the 21st Century' (2012) 90 Harvard Business Review 70.
[105] Davenport, Harris and Morision 83, stating, 'Segmentation in turn enables *differentiated action* – treating customers differently, or choosing the most efficient path in a flexible business process.' Differentiated action is consequently the progenitor of prescriptive action.
[106] Michael Walker, 'Prescriptive Analytics' (*Data Science Central*, 2013) <www.datasciencecentral.com/profiles/blogs/prescriptive-analytics> accessed 20 June 2019.
[107] Lustig and others.
[108] Ian Kerr and Jessica Earle, 'Prediction, Preemption, Presumption: How Big Data Threatens Big Picture Privacy' 66 Stanford Law Review Online 65, 67 regarding consequential and pre-emptive predictions.

processes and behaviours.[109] The same circular logic consequently exists. Embedded prescriptive processes, the sensorised infrastructure of the smart home, provide greater insight into key operational processes, such as the home's operation, and the focus of decision making thus shifts towards changing the behaviour of occupants to facilitate more efficient operational processes. At this stage, we reach the truly attentive home in Aldrich's framework, because decision making is now a machine rather than human-based activity built on the predicted inferences of patterns and behaviours that are balanced with the efficient operation of the home.

The heightened states of sensing and reasoning in the smart home, through the advent of dense sensorised collections and the enhancement of data analytic capabilities, have been a palpable driver for commercial entities that has led to a resurgence of the smart home. The commercial prospecting of smart home data is creating a range of new services and products that seek to offer a more individualised offering. The development of smart home insurance is a prime example of the actualities of commercialising smart home data – covered in much greater depth in Chapter 4. In the meantime, it is important to bring to light the underlying business logics that are increasingly pervading the smart home space, to consider the broader effect of commercialising all data in a truly collected world. Two aspects are important to consider: (1) the increasing drive for real-time, data-driven service provision and (2) the increasing automated mechanisms of prescription to 'nudge' behaviours.[110] Both are evident in the rapidly developing smart home insurance space and exemplify Lewis' final state – intervening.

3.5   Intervening: The Prescriptive Outcomes

The previous subsections have outlined the sensing apparatus of the smart home through a technological dissection of its anatomy and the data analytic processes that power predictive and prescriptive reasoning outputs. All that remains to be covered is the intervening process where predictive data outputs become part of prescriptive and real actions. The intervening phase of the smart home briefly detailed here regards the development of new smart home insurance business models,[111] as way of setting up coverage in Chapter 4 that closely examines how these models commercialise smart home sensor data.

---

[109] Davenport and Patil 71.

[110] Woodrow Hartzog, 'Opinions: The Case against Idealising Control' (2018) 4 European Data Protection Law Review 423, 427.

[111] Antonella Cappiello, *Technology and the Insurance Industry Re-configuring the Competitive Landscape* (Springer International Publishing 2018) 39.

During the past couple of years, rapid development of new, individualised insurance products based on the acquisition and analysis of real-time, sensorised data has taken place.[112] Smart home sensorised systems are now being utilised by insurers as a means of offering new products that are no longer based on traditional forms of pool insurance.[113] Instead, smart home–focused products and services increasingly rely on real-time, data-driven underwriting and pricing processes tailored specifically to individual circumstances.[114] The data used for risk profiling consequently shifts to the structured and unstructured sensorised collections of individual activities and behaviours[115] generated by the smart home and its devices.[116] Hence the smart home is one of the prime sites of data prospecting in the collected world.

However, even though these innovations are often trumpeted loudly, it is important to note that such developments are currently prospective and do not necessarily reflect the current state of smart home insurance developments, as detailed in Chapter 4. Nevertheless, the enhanced collection of data from sensorised devices is often presented as the imperative pathway for future smart home insurance business model opportunities.

Proponents argue that insurers could use the same type of sensorised data collections and analytic techniques utilised for car insurance purposes. The use of telematics techniques in the smart home would allow insurers to glean insights from sensor data to develop more accurate predictions of behaviour that are then used to 'risk profile' customers.[117] Prevention of damage and smart home data collections for insurance purposes thus have a synergistic effect.[118] Prevention, in the smart home

---

[112] Price Waterhouse Coopers, *Opportunities Await: How InsurTech Is Reshaping Insurance* (Global FinTech Survey, June 2016); Kai Riemer and others, *The Fintech Advantage: Harnessing Digital Technology, Keeping the Customer in Focus* (Capgemini & The University of Sydney Business School 2017); Tan Choon Yan, Paul Schulte and David Lee Kuo Chuen, 'InsurTech and FinTech: Banking and Insurance Enablement' in *Handbook of Blockchain, Digital Finance, and Inclusion* (Academic Press 2018).

[113] Sam Lewis, 'Insurtech: An Industry Ripe for Disruption' (2017) 1 The Georgetown Law Technology Review 491.

[114] Cognizant, *Next-Generation Insurance: Tapping Into the Intelligence of Smart Homes* (2015); Jay D Wilson, *Creating Strategic Value Through Financial Technology* (John Wiley & Sons 2017).

[115] Michelle Canaan, John Lucker and Bram Spector, *Opting In: Using IoT Connectivity to Drive Differentiation* (Deloitte University Press 2016); Markus Löffler and others, *Insurers Need to Plug into the Internet of Things – or Risk Falling Behind* (McKinsey & Company 2016).

[116] Yan, Schulte and Lee Kuo Chuen.

[117] Price Waterhouse Coopers, *Opportunities Await*.

[118] Martin Eling and Martin Lehmann, 'The Impact of Digitalization on the Insurance Value Chain and the Insurability of Risks' [2017] The Geneva Papers on Risk and Insurance – Issues and Practice.

sense, refers to property harm mitigation through different responsive sources.[119] These can include the provision of real-time information on home operations to homeowners for better self-monitoring processes or the development of automated professional monitoring services.

Insurers could use smart home data to better quantify household and individual risk and tailor customer premiums based on a highly individualised risk assessment. For example, certain actions – such as leaving smart doors or windows unlocked for long periods of time or failing to respond efficiently to app-related alerts – could be categorised as behaviour that would give rise to increased risks.[120] Those individual customers who take such actions deemed 'risky' would then be charged higher premiums based on their activities and inferred behaviours.[121] On the other hand, customers who are deemed to understand and mitigate their own risks – for example, those who keep doors and windows closed – could receive discounted premiums, including by the rather straightforward action of implementing certain smart home systems or products.[122]

It is also contended that smart home sensor data collections could be used to identify personality details or emotional characteristics that could be modelled for risk exposure. Again, examples are often put forward regarding the use of telematics and car insurance predicated on driver behaviour[123] that utilise sensorised data, including video footage. From the use of such data, insurers are able to detect complex cognitive states of drivers, including emotions such as frustration and fatigue.[124] The reasoning behind these models pertains to the correlation of certain emotional states that can be used to predict patterns of decision making that give rise to risk, such as impulsiveness or inattentiveness.[125] Similar strategies could equally be employed in the smart home[126] to

[119] Canaan, Lucker and Spector.
[120] Stacey Higginbotham, 'Why Insurance Companies Want to Subsidize Your Smart Home' (*MIT Technology Review*, 2016) <www.technologyreview.com/s/602532/why-insurance-companies-want-to-subsidize-your-smart-home/> accessed 20 June 2019.
[121] Dominic Cortis and others, 'InsurTech' in Theo Lynn and others (eds), *Disrupting Finance: FinTech and Strategy in the 21st Century* (Springer International Publishing 2019) 75.
[122] Canaan, Lucker and Spector.
[123] Liz McFall and Liz Moor, 'Who, or What, Is Insurtech Personalizing? Persons, Prices and the Historical Classifications of Risk' (2018) 19 Distinktion: Journal of Social Theory 193, 196–7.
[124] Rana el Kaliouby, 'We Need Computers with Empathy' (*MIT Technology Review*, 2017) <www.technologyreview.com/s/609071/we-need-computers-with-empathy/> accessed 20 June 2019; Laurence Goadsuff, 'Emotion AI Will Personalize Interactions' (Gartner 2018) <www.gartner.com/smarterwithgartner/emotion-ai-will-personalize-interactions/> accessed 20 June 2019.
[125] Canaan, Lucker and Spector.
[126] Eling and Lehmann.

develop usage-based home insurance schemes. In these models, an insurer would collect and analyse smart home sensor data to monitor individual behaviours and adjust premiums in real time based on activities and behaviours.[127] The extension of telematics to smart home insurance is therefore considered by some to be logical.[128]

Smart home sensor data, particularly generated by smart home systems, are seen as a key component for the development of new business models, because they allow insurers to identify, monitor and prevent risks emerging in the home which might otherwise lead to a future claim.[129] However, an insurer's ability to mitigate risk with sensorised devices goes beyond merely identifying latent problems in the home and notifying their existence.[130] Data-driven insights can also be used to 'nudge' policyholders towards more positive behaviours that are less risky and are less likely to result in claims arising in the first place.[131] Nudges are effectively Lewis' smart home interventions and the outcomes of prescriptive analytic processes.[132]

Again, examples of sensorised behavioural nudging are starting to come to fruition in the telematics and car insurance domain.[133] Smartphones are increasingly being used to monitor and profile driving behaviours. Similarly, the 'black box' equivalent of modern cars can now generate significant data on driver behaviours such as speeding in certain locations, braking techniques and phone or other media usage while driving.[134] The prescriptive nudge then involves information disclosure mechanisms such as providing dashboard visualisations for drivers or sending automated reminders about questionable driving behaviours. Individual driving habits could therefore be improved, which benefits both the insured, the insurer and society at large.[135]

---

[127] Higginbotham.
[128] Wilson.
[129] Price Waterhouse Coopers, *Insurance's New Normal: Driving Innovation with InsurTech'* (2016).
[130] Yan, Schulte and Lee Kuo Chuen.
[131] Michael Naylor, *Insurance Transformed* (Springer 2017).
[132] Karen Yeung, '"Hypernudge": Big Data as a Mode of Regulation by Design' (2017) 20 Information, Communication & Society 118 for a comprehensive overview of key issues regarding the use of nudging structures in automated systems.
[133] Doug Drinkwater, '10 Real-Life Examples of IoT in Insurance' (*Internet of Business*, 2016) <https://internetofbusiness.com/10-examples-iot-insurance/> accessed 20 June 2019.
[134] Canaan, Lucker and Spector.
[135] Richard Clarke and Ari Libarikian, 'Unleashing the Value of Advanced Analytics in Insurance' (McKinsey & Company, 2014) <www.mckinsey.com/industries/financial-services/our-insights/unleashing-the-value-of-advanced-analytics-in-insurance> accessed 20 June 2019.

It is therefore possible that similar prescriptive strategies will develop with smart home insurance, which means that data collection will be needed for a whole range of home-based micro activities, such as checking the battery levels of important devices like smoke alarms.[136] However, and perhaps not surprisingly, these types of developments have also raised concerns regarding customer control and autonomy, as outlined in later chapters.

3.6   Conclusion

The convergence of different technological frameworks in the smart home has created new opportunities for commercialising smart home sensor data. Sensors are important in this regard, as they provide the means and opportunity to collect fine-grained data about individual patterns of behaviour. These new data collection prospects are spurring new business logics that aim to provide increasingly individualised ser-vice delivery on a progressively real-time basis. Sensorised spaces, such as the smart home, are giving rise to new business models made possi-ble by infrastructural components that are themselves being sensorised to fulfil data collection requirements. A circular logic is unfolding: sen-sors provide, in ever greater volumes, quality data which are utilised by business models that generate new value propositions for infrastructure or device manufacturers based on sensorisation. More sensors mean more data. More data means new value opportunities. The prospect of new value means more sensors. All of this fuels the seemingly inexo-rable commercial impetus for the collected world.[137]

However, as noted previously, that impetus must currently work around the fragmented technological framework of spaces like the smart home. The effect, as outlined in Chapter 4, is that different smart home business models utilise sensor data in different ways depending on their data generation and collection capabilities.

Another important factor needs to be noted before Chapter 4 is read and Part III's later analysis is considered. At present, there is no single dominant smart home data collection business model. However, that may well change with the increasing advent of Google Home and Amazon Alexa. Amazon's and Google's dominance of the smart home

---

[136] Canaan, Lucker and Spector.
[137] Helen Nissenbaum, 'Must Privacy Give Way to Use Regulation?' 2015–16 Faculty Seminar: Democracy, Citizenship, and Constitutionalism <www.sas.upenn.edu/andrea-mitchell-center/sites/www.sas.upenn.edu.dcc/files/Nissenbaum-UPenn-Democracy.pdf> accessed 14 May 2019, 9.

market is by no means confirmed, given the fragmented and chaotic infrastructural environment of the smart home. Nevertheless, two considerations are clear. First, smart home sensor data appear to be key to business model ambitions. Second, the platform titans, and other major manufacturers, are seeking to be major players in the newly emerging smart home data collection markets.

Chapter 4 now delves into these considerations more comprehensively when we examine commercialisation strategies of sensor data involving smart home insurance business models and their data collection requirements.

4   Commercialising the Collected

4.1   Introduction

The previous chapters outline the scope of the collected world and overview the acquisition of data from the smart home as a target of significant commercial interest. In this chapter, we delve further into smart home considerations and provide a fine-grained examination of evolving commercialisation strategies that entail sensorised data collections from the home. The chapter thus highlights the hugely important role that personal information plays in the development of new smart home–focused business models. Moreover, and more importantly, it provides an opportunity to bring to the surface the often-imperceptible logics that pervade the evolution of new data accumulation models. The case studies covered in this chapter will also be used later in the book to consider the varying influences of existing information privacy regulatory regimes and the challenges posed to those frameworks from ubiquitous forms of sensorised data collection.

Chapter 3 introduced some of the broader logics underpinning developments in smart home insurance. Building on these developments, this chapter examines three smart home insurance products with specific focus on the conceptualisation of emerging data collection partnership models. Each model involves a smart home automation system or device – namely, Fibraro, Canary and Nest, respectively. Each smart home system operates within the same component architecture outlined in Chapter 3 and collects similar types of sensorised data. However, the type of data-focused partnering arrangement appears to be different in each model, which has an important impact on the data collection and analysis strategies employed.

In turn, these different partnership models give rise to some common information privacy considerations but also to some unique considerations pertinent to each model, as Chapter 7 outlines. More importantly, much like the technological anatomy of the smart home, the detailed examination of data collection capacities and data exchange processes reveal a dense network of partnership arrangement that stretches beyond the confines of any given smart home. The partnership arrangements thus highlight the complexity of the collected world and the challenges that arise for information privacy law from commercialisation of sensorised data, particularly from the smart home.

The smart home insurance products examined in this chapter are part of the broader use of Insurance Technology (InsurTech) developments currently unfolding at a rapid rate.[1] Total global funding for InsurTech investments amounted to $1.42 billion across eighty-five agreements in the first quarter of 2019.[2] InsurTech refers to the use of different technologies, but it also reflects different types of partnership arrangements that aim to provide new service benefits for customers.[3] Established insurers are now viewing technological developments, such as IoT, as potential threats to their traditional business model and as a response are seeking to partner with new start-up partners.[4] The partnership model appears to be the dominant platform for InsurTech expansion[5] which is driving unprecedented levels of investment.[6]

Both sides of these new partnership arrangements believe that they are mutually beneficial and will improve service provision for customers.[7] Traditional insurers have experiences of working within sophisticated regulatory regimes, and start-ups are believed to have greater levels of technological impact, particularly regarding the use of sensorised technologies.[8] InsurTech investment and partnering strategies are also inherently linked to smart home developments, as a number of the targeted InsurTech start-ups provide smart home automation services. As this chapter details, smart home automation or control systems are significant partners of interest for established insurance companies seeking to expand their technological ambits in relation to home insurance services or products.

---

[1] Bernardo Nicoletti, *The Future of Fintech: Integrating Finance and Technology in Financial Services* (Palgrave Macmillan 2017).

[2] 'InsurTech Investment Records Smashed in Q1, 2019' (*MarketWatch*, 2019) <www.marketwatch.com/press-release/insurtech-investment-records-smashed-in-q1-2019-pricing-tools-in-the-spotlight-2019-05-09?dist=bigcharts> accessed 20 June 2019.

[3] Price Waterhouse Coopers, *InsurTech: A Force for Good* (2016).

[4] Ibid.

[5] OECD, *Technology and Innovation in the Insurance Sector* (OECD 2017).

[6] Price Waterhouse Coopers, *Insurance's New Normal*.

[7] Ibid.

[8] Martin Blake, 'Insurtech: An Industry Dilemma' (*KPMG*, 2017) <https://home.kpmg.com/au/en/home/insights/2017/03/insurtech-dilemma-insurance-technology.html> accessed 20 June 2019 Wilson.

In the academic literature, it appears that InsurTech developments have generally been considered part of a broader spectrum of FinTech,[9] even though InsurTech and FinTech innovations have different emphases. InsurTech is concerned with individual customer service improvements, whereas FinTech generally refers to business innovations.[10] InsurTech developments in the scant academic literature are thus often examined under the broader ambit of FinTech,[11] digital insurance,[12] digital finance[13] or digital marketplaces.[14] That is not necessarily surprising, given the relative recentness of InsurTech developments. Equally unsurprising is that the InsurTech environment has been referred to as vast, heterogeneous and opaque and that it attempts to categorise environmental complexities as unidimensional.

Given the newness of this topic and the limited coverage available, this chapter begins by providing an evolutionary overview of smart home insurance developments to situate the analysis of the three partnership arrangements that will be described.

4.2   The Emergence of Smart Home Insurance

Chapter 3 details that the smart home commercial market is a fast-moving area. It is also fragmented due to the many new home automation systems and sensorised home devices that have come onto the market in a short space of time.[15] The lack of a sustainable interoperability protocol, or framework, adds further complexity to the data collection possibilities of smart home insurance models.[16] Thus, the sheer volume of relevant actors involved in smart home collections – such as telecom operators, utility management companies and technology providers – makes standard forms of data collection difficult.

---

[9] Harmut Mai, 'Preface: Fin- & Insuretech' in Claudia Linnhoff-Popien, Ralf Schneider and Michael Zaddach (eds), *Digital Marketplaces Unleashed* (Springer 2017).

[10] OECD, *Technology and Innovation in the Insurance Sector* 8.

[11] Mai 331.

[12] Michael Cebulsky and others, 'The Digital Insurance: Facing Customer Expectation in a Rapidly Changing World' in Claudia Linnhoff-Popien, Ralf Schneider and Michael Zaddach (eds), *Digital Marketplaces Unleashed* (Springer 2017).

[13] Yan et al., *Handbook of Blockchain, Digital Finance, and Inclusion*.

[14] Claudia Linnhoff-Popien, Ralf Schneider and Michael Zaddach, *Digital Marketplaces Unleashed* (Springer 2017).

[15] Emily Signer, 'How Smart Homes Can Keep You Safe – And Save You Money' (*Gear Patrol*, 2017) <https://gearpatrol.com/2017/06/08/smart-home-insurance-discount/> accessed 20 June 2019.

[16] Munich RE and Hartford Steam Boiler, *Connected Home Technologies* (*Gear Patrol*, 2016). <https://gearpatrol.com/2017/06/08/smart-home-insurance-discount/>.

As things currently stand, it is challenging for any business seeking to collect all data from a smart home because of the volume of data collection from different smart home devices.[17] However, as Chapter 3 alludes to, that situation could change due to the widespread implementation of popular voice control systems, such as Amazon Echo and Google Home, that could be used to develop a dominant data collection model. Such a model will generate significant information privacy concerns that are redolent of the collected issues highlighted throughout this book. Before we get that far, it is necessary to provide a brief overview of smart home insurance developments to situate the chapter's later analysis. Three phases of business model development are pertinent: mutual product promotion, smart home monitoring and prevention, and data acquisition partnerships.

4.2.1   Mutual Product Promotion

The first phase of smart home InsurTech developments began around 2013 with mutual product promotion arrangements. These developments came from an established footing involving previously discounted insurance schemes for the implementation of physical security products. Established insurers began to partner with smart home service providers to provide discounts and targeted information to customers on both sides of the arrangement. Accordingly, the insurers' customers were informed about a specific smart home service or product and vice versa. Premium discounts were offered to new and existing customers based on the implementation and use of the smart home service. The insurer benefits from the implementation of such devices because it is likely to enable enhanced protection and prevention strategies in the home.[18] Simply put, homeowners are more aware of insurable risks.[19] The smart home service also benefits from the discounted insurance promotions, as it provides an added incentive for insured homeowners to purchase and adopt a specific type of smart home service or product.

In 2013, one of the first promotional models opened in the UK with the partnership between the insurer Axa and British Gas. The partnership was based on the latter's home heating system, Hive, which allowed users to remotely control home heating systems by mobile app.[20] In this

---

[17] Cognizant 10.
[18] Yan, Schulte and Lee Kuo Chuen 264.
[19] Cognizant 7.
[20] Freddie Roberts, '10 Leading Partnerships Driving IoT Adoption in Insurance' (*Internet of Business*, 2017) <https://internetofbusiness.com/10-iot-companies-insurance/> accessed 20 June 2019.

partnership, customers who purchased a Hive product online were auto-matically redirected to Axa's website to receive information on home insurance policies.[21] At inception, the promotion arrangement was purely based on information provision but has now developed into a combined partnership that provides British Gas home and contents insurance.[22]

During the 2013 period, similar US partnerships developed that involved the insurance companies Liberty Mutual and State Farm. In December 2013, Liberty Mutual began partnership with the smart home system Vivint, with a promotion discount package.[23] Each part-ner's customers could contact the other partner to receive either a discount on premiums or on the purchase of smart home products. Liberty Mutual significantly expanded its partnership portfolio over the next couple of years to include a range of different smart home product providers, including Nest and Canary.[24] The US insurer State Farm also began a similar promotional and discount strategy with Canary[25] and with the leading security system ADT.[26] Similar arrange-ments also involved the insurer American Family Insurance and the smart doorbell producer Ring.[27] In Europe, Axa France embarked on a promotional partnership with the smart light bulb Phillips Hue to mitigate potential theft risks.[28]

---

[21] CB Insights, *How Major Insurers Are Teaming Up with Internet of Things Companies in One Infographic* (2015).

[22] 'Home Insurance' (*British Gas*, 2018) <www.britishgas.co.uk/home-services/home-insurance.html> accessed 20 June 2019.

[23] Svetlana Volosovich, 'Insurtech: Challenges and Development Perspectives' (2016) 3 International Journal of Innovative Technologies in Economy 39. See also 'Home Automation Giant Vivint Partners with Liberty Mutual Insurance to Offer Its Customers Savings on Auto and Home Insurance' (*Vivint.SmartHome*, 2013) <www.vivint.com/company/newsroom/press/Home-Automation-Giant-Vivint-Partners-with-Liberty-Mutual-Insurance-to-Offer-its-Customers-Savings-on-Auto-and-Home-Insurance> accessed 20 June 2019.

[24] 'Canary Announces Partnership with Liberty Mutual Insurance' (*Market Wired*, 2016) <www.marketwired.com/press-release/canary-announces-partnership-with-liberty-mutual-insurance-2141220.htm> accessed 20 June 2019.

[25] 'State Farm Teams Up with Canary to Bring Customers Exclusive Offers' (*CISION PR Newswire*, 2015) <www.prnewswire.com/news-releases/state-farm-teams-up-with-canary-to-bring-customers-exclusive-offers-300107774.html> accessed 20 June 2019.

[26] Shawn Kim, 'Which Insurance Companies Offer Discounts If You Have a Security System?' (2016) <www.homesecuritylist.com/which-insurance-companies-offer-dis-counts-if-you-have-a-security-system/> accessed 20 June 2019.

[27] Drinkwater. Ring was recently purchased by Amazon and is now part of Amazon's Echo-based smart home family.

[28] 'Philips Hue Partners with American and European Telecom, Insurance and Energy Companies on Smart Home Solutions' (*LEDInside*, 2016) <www.ledinside.com/news/2016/3/philips_hue_partners_with_american_an_european_telecom_insur-ance_and_energy_company> accessed 20 June 2019.

The promotional model is a developmental extension of insurer discounts and provision of information for physical security products and smart home products. Importantly, for this chapter, the partnership arrangement of both the insurer and the smart home service or smart home device maker is not yet based upon the collection of sensorised data from a smart home system or devices. Instead, the arrangement regarded a process of bilateral promotion and information offerings between two parties to serve a mutually recognised interest. The next generation of smart home insurance developments saw the beginning of data-focused exchange strategies.

4.2.2   Smart Home Monitoring and Prevention Services

The second phase of partnership development, the advent of smart home–focused monitoring and prevention services, saw a specific focus of sensorised smart home data for commercial purposes. The second phase is important because it placed greater attention on the data collected by smart home service providers and how that data could be better utilised by insurer partners for key business development, including risk assessment, new product lines and pricing.[29] However, while the second phase introduced new sensorised data acquisition prospects, the basis of developments tended to be focused upon the implementation of better monitoring and prevention systems.

Other monitoring and prevention partnerships offered a different approach, especially where the monitoring of the home was undertaken by the smart home service rather than the homeowner. An example is the partnership between Allianz and Panasonic. In 2015, a partnership arrangement was formed to launch a home automation solution that integrated insurance, customer assistance and home automation technology to protect customer insured property.[30] The Panasonic Smart Home & Allianz Assist service connected Panasonic's Smart Home monitoring and control system with home protection services provided by Allianz Global Assistance, the assistance subsidiary of Allianz Worldwide Partners.[31] In the event of an incident occurring in a

---

[29] Löffler and others.

[30] Maryvonne Gray, 'Leading Player Allianz Partners with Tech Giant' (2015) <www.insurancebusinessmag.com/au/news/breaking-news/leading-player-allianz-partners-with-tech-giant-56780.aspx>.

[31] Maryvonne Gray, 'Leading Player Allianz Partners with Tech Giant' (*Insurance Business Australia*, 2015) <www.insurancebusinessmag.com/au/news/breaking-news/leading-player-allianz-partners-with-tech-giant-56780.aspx> accessed 20 June 2019.

protected home, the Panasonic system sent a sensor alert to a customer's smartphone or tablet via an app. An indoor siren was also automatically activated, and notification was given to the Allianz Global Assistance service centre. The Panasonic system consisted of a set of intelligent devices that included window and door motion sensors, water leak sensors and breaking-glass sensors and the indoor siren. These were connected via a secure and proprietary Ultra Low Energy (ULE) standard to a dedicated Panasonic hub located in the home or property.[32]

The partnership between UK start-up insurer Neos and the smart home automation system provider Fibraro also had a monitoring and prevention purpose. The Neos smart home insurance system required the implementation of several sensorised devices in the home.[33] These devices included motion detectors and security cameras for self-monitoring purposes, similar to the developments already mentioned, and a flood sensor that could detect potential water leaks and a valve that could proactively turn off water if a leak was detected.[34]

The Neos model consequently relied on a combination of Neos-generated automated remote monitoring of collected sensorised data and updated reporting to customers. Customer reporting could include enhanced and near-real-time notification of potential incidents that could be escalated to customer warnings that required attention.[35] Under the Neos model, the customer was at the front line because, as first responder, they are often in the best position to identify false alarms and real incidents.[36] The Neos system was also moving towards post-incident connection and supply of tradespersons required to fix household incidents quickly.[37]

The development of increasingly automated professional and customer-oriented monitoring is important, because it bridges the initial promotional developments of the first phase to the more advanced

[32] Ibid.
[33] The web page for this citation has subsequently been removed. A PDF version is on file with the author. 'Meet Our Smart Home Security Devices' (*Neos*, 2017) <https://neos.co.uk/meet-our-smart-home-security-devices/> accessed 20 June 2019.
[34] Jay Borkakoti, 'Escape of Water Claims, Smart Sensors and the Role of Insurance' (*Insurance Insights*, 2017) <http://blogs.lexisnexis.com/insurance-insights/2017/02/escape-of-water-claims-smart-sensors-and-the-role-of-insurance/> accessed 20 June 2019.
[35] 'How Neos Plans to Combine Insurance with the Smart Home' (*Andrew Lucas London*, 2016) <www.andrew-lucas.com/journal/neos-plan-insurance-smart-home/> accessed 20 June 2019.
[36] Ibid.
[37] 'Neos Takes New Approach to IoT with Smart Insurance' (*Andrew Lucas London*, 2016) <www.andrew-lucas.com/journal/neos-heralds-new-approach-iot-smart-technology-insurance-package/> accessed 20 June 2019.

data collection aspirations of smart home–focused insurance partnerships. Sensorised data acquisition strategies come directly to the fore in the next phase of development.

4.2.3   Data Acquisition Partnerships

Under this type of partnership arrangement, insurers partner with smart home service providers in a deeper and more integrated fashion regarding the acquisition and analysis of sensorised device data for risk-based pricing and for the development of new products and services. The first developments of this type took place in the car insurance space. In 2014, the US car insurer Progressive partnered with Zubie, a start-up producer of sensorised devices to assist drivers in monitoring their driving performance.[38] Zubie customers shared data generated by the device with Progressive to earn a discount from Progressive's Snapshot programme.[39] In a similar venture, Progressive partnered with Censio, maker of a sensorised device that monitors driver and in-car smartphone usage, an arrangement that also involved customer data exchange from Censio to Progressive.[40]

These telematics developments are important because they outline a model of customer-focused data exchange that is adopted in the smart home insurance sphere. In doing so, it becomes possible to track the development of sensorised data exchange that is now being incorporated into the promotional-type partnering arrangements just detailed. Examples of this move involve Liberty Mutual and State Farm and their ongoing partnership with the smart home security provider Canary. In 2016, it was reported that Canary, in conjunction with said insurance partners, was preparing a pilot program that would allow customers to share select, limited information with the two insurance companies.[41] The information was then used for more accurate insurance pricing that would provide Canary customers with better priced, individualised offers.[42]

During the same year, Neos became the first insurance company to launch a dedicated smart home insurance service[43] that used sensorised

---

[38] Drinkwater.
[39] Roberts.
[40] CB Insights.
[41] Nathan Golia, 'Liberty Mutual Partners with Smart Home Company Canary' (2016) <www.dig-in.com/news/liberty-mutual-partners-with-smart-home-company-canary> accessed 20 June 2019.
[42] 'Canary Announces Partnership with Liberty Mutual Insurance'.
[43] 'Neos Takes New Approach to IoT with Smart Insurance'; 'How Neos Plans to Combine Insurance with the Smart Home'.

data collections from a smart home automation system.[44] The Neos model was based on a partnership arrangement with the home automation system provider Fibraro.[45] A customer was provided with a Fibraro hub and a set of sensorised devices for installation in the home. The provided devices varied upon what Neos package was purchased, and the more expensive packages provided more sensorised devices.

Each sensorised device was then linked to the Fibraro hub, and the customer could activate supplied devices through a separate Fibraro controller or by smartphone via the Neos app.[46] The model was designed to reduce the need for insurance claims, as it provided customers with enhanced real-time knowledge about how the home operated, thus permitting potential problems to be prevented before they arose. The sensorised device data also provided Neos with a significant amount of data that could be used to better anticipate customer needs and behaviours, as the analysis that follows outlines.

Neos owned the devices and collected them upon the cancellation of an insurance contract. At the time of writing, the devices were not customisable by the customer and additional devices could not be added to the Neos-managed Fibraro hub. However, Neos was seeking to add a customisable function to its system as well as the capability to include certain additional devices beyond those provided, such as Netatmo cameras and other Fibraro products. Neos also added Amazon Alexa integration, which can be used as an alternative to the Neos app. Going back to Chapter 3 and the anatomy of the smart home, the fragmentation of smart home control started to become evident in the construction of business models, as exemplified by Neos' response to wider-scale Echo usage.

The Neos/Fibraro model is an important development, as it signified the first concentrated attempt to utilise a smart home system for purposeful acquisition of sensorised data for insurance purposes. The Neos model started to action the actual analysis of smart home sensorised data as a fundamental part of its business model, both as a prevention service and as a provider of reduced insurance premiums, based specifically on the data generated from smart home devices.

Building on the Neos developments, the first smart home usage-based insurance partnership now appears to be in operation between

[44] Amelia Heathman, 'UK Startup Neos Is Combining IoT Devices with Insurance to Make Your Connected Home Safer' (2017) <www.verdict.co.uk/neos-startup-iot-insurance/> accessed 20 June 2019.
[45] 'Terms and Conditions' (*Neos*, 2017) <https://neos.co.uk/terms-and-conditions/> accessed 20 June 2019.
[46] Ibid.

Local Tapiola, Finland's largest mutual insurance company, and smart home system Cozify. The partnership is like Neos/Fibraro and forms the foundation for Finland's first smart home insurance based on the findings of an initial pilot in 2016, which identified a customer demand for 'proactive companion' insurance.[47] Similar to the Neos model, the Local Tapiola partnership utilises sensorised data from Cozify's home automation system to provide a customer notification, prevention system via a smartphone app. However, the partnership also appears to be considering a significant shift to usage-based insurance derived from sensorised smart home device data. The Cozify hub would therefore be used as a telematics equivalent to specifically 'enable learning about the factors that affect risk profiles and to provide additional value to customers'.[48] Expansive near-real-time data sharing with partner systems also appears to be contemplated.[49]

4.3   Smart Home Data Exchange Partnerships

The developmental phases of smart home insurance give rise to different data acquisition considerations. This section builds upon the preceding coverage, and the component technical infrastructure identified in Chapter 3, to elucidate three conceptualised models of sensorised data acquisition involving insurer and smart home system or device provider. Each model is predicated upon sensorised data collection, but the partnership relationships and the use of collected smart home data are different. The differences are evident from an analysis of publicly available privacy policies of the main partnerships and other documentation.[50] The examination of privacy policies revealed the types of data collected, the collectors' classification of personal information and data exchange arrangements, including which data was shared and for what purpose.

It should also be noted that while these models are different, they are not distinct, and there can be movement between models.

---

[47] Cozify, 'The First Smart Home Insurance in Scandinavia' (2017) <http://blog.cozify.fi/2017/09/the-first-smart-home-insurance-in.html> accessed 20 June 2019.

[48] Ibid.

[49] Ibid.

[50] The analysis of publicly available privacy policies was conducted in the first quarter of 2018. Since then, some have been updated due to the implementation of the GDPR. The analysed privacy policies are referred to in this chapter, and details of the updated ones are referenced accordingly. Changes to updated privacy policies were minimal, in terms of the research focus, and thus the analysed 2018 privacy policies are referenced as the basis of research. Copies of all relevant privacy policies are on file with the author.

Table 4.1. *Insurer and smart home data partnership models*

Model No.	Model Type	Relationship Type	Data Focus of Partnership	Insurer/Smart Home Partners
1	Partnered Data Acquisition	Monogamous; one-to-one	Collection	Neos/Fibraro; local Tapiola/ Cozify
2	Partnered Intermediary	Polygamous; many-to-many	Collection; connection	Liberty Mutual,[51] All State/Canary
3	Platform Entity	Dominant polygamous; one-to-many	Collection; connection; condition creation	Liberty Mutual, American Family/Nest (Google)

This a reflection of the fluid environment of smart home insurance. Nevertheless, these models provide a fine-grained insight into how collected smart home data is used for commercialisation purposes. More importantly, as this chapter alludes to and is covered in greater depth in later chapters, the different models have assorted information privacy law implications, particularly in relation to the categorisation and prospective uses of personal information.

As such, the examination of the three models illuminates the application of information privacy law protections and the infrastructural consequences of a collected world. What can be reiterated, much like with the smart home considerations identified in Chapter 3 and the collected world envisaged in Chapter 2, is that this is a complex, rapidly changing and uncertain area of prospective involvement, across several different frames. Table 4.1 outlines the different partnership relationship types and the primary focus of data operations in each partnership model. The operational data structures, the data collected and an exploration of each model's characteristics is detailed in the table.

4.3.1 The Partnered Data Acquisition Model

The first model of data operation is entitled Partnered Data Acquisition, which features a one-to-one partnership model between an insurer and a smart home system provider. Data generated from the customer's smart home system is exclusively collected for and used by the insurer to either provide an enhanced home protection service (e.g. Neos/Fibraro) or develop new usage-based smart home insurance predicated on more

[51] As demonstrated in the foregoing coverage, Canary was a technical partner with Liberty Mutual. The status of this commercial relationship, as of May 2019, is unclear, as Canary has removed Liberty Mutual as a partner insurer.

accurate risk profiling of individual behaviours (e.g. Local Tapiola/ Cozify). In both instances, sensorised data collection is at the heart of the business model. There is an intimate link between the sensorised device; the component smart home infrastructure; and the ultimate, prescriptive output, such as a customer notification or warning in a prevention system or an adjusted premium in a pricing system.

For the most part, the customer remains in control of home automation decisions and is encouraged by the partnership to retain control. Customer control retention provides significantly more data about how the individual customer uses the smart home components, which is captured through sensorised data collection. Customer control data can then be used to infer activities and behaviours in the home which can then be factored into algorithmic outputs regarding intervention or pricing. As such, going back to the coverage of the smart home's infrastructural anatomy, there is a direct link between the use of controllers by individuals, analysis of predicted behaviours and the ultimate prescriptive output. These outputs form the attentive customer notifications of Neos, where the home automation system can differentiate between false alarms and real events based on analysis of historical and repetitive usage patterns.[52] Similarly, inferences of customer activities and behaviours can be used for usage-based risk profiling for smart home insurance purposes, such as Local Tapiola/ Cozify. The operational data structure of this model is therefore vital to understand, given the direct link between user agency, data collection and the predicted or prescriptive output as the core foundation of the business model.

4.3.1.1 Operational Data Structure The insurer is the primary data collector in the Partnered Data Acquisition model, as it collects smart home sensor data directly from the insured. The role of the smart home system is as data collection agent or supplier of smart home systems that enable collection. For example, Neos had a procurement partnership with Fibraro for its home automation system.[53] Fibraro supplied Neos with the hub and a range of dedicated sensorised devices for installation

---

[52] 'FAQs' (*Neos*, 2017) <https://neos.co.uk/faqs/> accessed 20 June 2019.
[53] Note, however, that the exclusive partnership between Neos and Fibraro no longer seems to be in operation due to Fibraro's business decision to pull out of the UK market. Neos still provides the same range of sensorised products, but it is unclear whether there is now an explicit and exclusive partnership arrangement with a different smart home system.

in the insured home. These included the following, depending upon what package was purchased:

- 2 Fibraro wireless motion sensors
- 1 Netatmo indoor wireless camera
- 2 Fibraro leak detectors
- 2 Fibraro wireless smoke sensors
- 2 Fibraro window/door sensors

The sensorised devices continuously collect data that can then be analysed to detect when an event is occurring or likely to occur. For example, if the smoke detector senses smoke or the motion sensor identified movement in its immediate surroundings, then that can be used as a customer alert. All three states of Lewis' smart home (as Chapter 3 outlines) – sensing, reasoning and intervening – are thus employed. The data is sent from one of these sensors, using the low-power Z-Wave protocol, to the Neos hub. The Neos hub subsequently translates the Z-Wave signal to a Wi-Fi signal. The sensor's event data is then sent to Neos' cloud-based servers over the Internet. Likewise, data from a customer's mobile phone usage would also be sent to Neos' servers. As highlighted in Chapter 3, some of the sensorised devices, most notably the indoor camera, must connect to the hub directly via Wi-Fi due to the battery requirements and data exchange size of the camera and its outputs.

It should also be noted that the exact process of data collection will vary under this model, as it will depend upon the devices and protocols that the insurer/home automation system employs. The Neos/Fibraro systems required a hub because it used sensors which rely on the Z-Wave protocol for communicating sensor data. A hub is therefore required to translate the low-power signal into a communications protocol that can also connect to phones, the Internet or the Cloud via Wi-Fi. Canary, for example, does not rely on Z-Wave or other smart home sensor data exchange protocols and connects directly via Wi-Fi. This point is seemingly small, but it could have significant implications for sensorised data collection, because a more expansive system, which can connect more devices, will collect more data regarding individual usage and control in the home. Equally, the location of the sensorised device may impact upon the volume and quality of sensorised data collected.

4.3.1.2 Data Collected   Neos collects a variety of different types of data in relation to its smart home insurance. Some of the data collections are standard, such as the collection of personal information upon registration. However, and importantly, a range of other sensorised data is collected which is directly relevant to its smart home insurance model.[54] These include

- event data collected from installed sensorised devices,
- customer mobile phone usage data related to the use of the Neos app and
- customer configuration data based on the configuration settings of a customer's mobile device.[55]

Table 4.2 identifies the wide range of data that is collected by Neos from the Fibraro system. It should also be noted that all this data is classified as personal data by Neos, thus subject to data protection or information privacy law. Some personal data is collected automatically and specifically from the customer, such as registration details. Other automatic, always-on collections from sensorised devices also takes place. Finally, Neos customers also provide opt-in permissions in relation to certain specified types of data, particularly in relation to mobile phone settings and social media profiles.

4.3.1.3 Model Characteristics   Several characteristic observations about the Partnered Data Acquisition are applicable. As exemplified by the Neos/Fibraro partnership, a significant amount of personal information is collected under that model. The reason for the volume of data collection is likely the direct link between sensorised data acquisition and Neos' core business model of preventative smart home insurance. Neos' product is dependent upon sensorised collections to identify potential environmental risks in the home and to better tailor customer notifications to individual behaviours. Local Tapiola's model with Cozify, similarly, will require the same type of collections. To do this, a significant amount of data needs to be collected from the smart home component infrastructure, including the primary controller – a customer's smartphone. The behavioural inferences that can be gained from smartphone use analysis may indicate why Neos is placing such great emphasis on the collection of customer smartphone data.

The data exchange relationships are also binary and straightforward. The Fibraro system was essentially a collection agent for Neos. There are currently no significant additional data exchange or data sharing arrangements in place. All data is collected directly by Neos by way of

---

[54] Neos, 'Privacy Policy' (2017) <https://neos.co.uk/privacy-policy/> accessed 20 June 2019.
[55] Ibid.

Table 4.2. *Smart home data collected by Neos*

Data Type	Details of Data Collected	a. Automatic/Opt In	b. Comment
Registration	Personal details such as name, username, email, address, etc.	Automatic	
Event data	Continuous collections from installed sensorised devices; video exception	Automatic for most sensorised devices, e.g. motion detectors; opt-in controls around video collection and sharing	
Mobile phone usage re app	Neos states that data collected 'includes but [is] not limited to' traffic data, location data, logs, error and crash reporting, and other communication data and the resources that you access.	Automatic	
Mobile device usage or configuration	Presence of other applications, unique device-identifying information and any cellular network, IP, Wi-Fi or Bluetooth data	Automatic	
Mobile device permissioned access	Address book, photos, geolocation, gyroscopes, data from your cameras or microphones	Opt in	Neos states, 'You may choose not to give permission to share this data, but it may restrict the usage of certain features of the mobile application.'
Social media profiles	Information from any social network or other online account	Opt in	
Browsing patterns	IP address, operating system and browser type	Automatic	Used for system information and 'to report aggregate information to advertisers'

sensor technology that they have procured from Fibraro. Accordingly, this model could be characterised as *monogamous* in nature, as it is a one-to-one system of data exchange involving two principal parties.

The system is also 'closed' in the sense that only data from the Fibraro system can be collected. However, as noted earlier, it appears that the Neos model is fluid in its development, as Neos is seeking to expand its range of technical and insurance partners. This could also be influenced by the fact Fibraro has stopped selling its products direct to the UK consumer market. If Neos does continue to evolve as suggested, then this could have important data collection consequences, because it could shift the Neos system from one characterised as 'closed' to one that is increasingly 'open' as a new range of additional sensorised devices from other smart home product manufacturers can be connected to the Neos system by the customer. At the same time, the leveraging of different customer device usage may also make it more attractive to other potential technical partners, thus shifting the primary role of data partnership from that of acquisition agent to exchange intermediary.

4.3.2   Partnered Intermediary

The second model, Partnered Intermediary, is more complex than the Partnered Data Acquisition model in several ways. First, while the second model still involves individual partnerships – for example, Liberty Mutual/Canary – these partnerships are not exclusively monogamous. An insurer can have multiple relationships with different smart home automation systems or services and vice versa. The primary reason for this relationship appears to be the role that each partner has in the context of data operations. Under the Partnered Intermediary model, the primary purpose of smart home partnership is not purely predicated upon data collection agency.

Both insurer and smart home systems have data collection roles, but these roles are enmeshed with a broader focus on data exchange of customer data for mutual benefit. The focus of data operations is not solely about the collection of sensorised data from smart home systems and devices. Rather, it is about the collection and exchange of enough customer data to facilitate *connection* between multiple partners for the purposes of product discount and promotion. The role of the partnered parties is thus that of an intermediary that collects customer data for exchange strategies to fulfil better connections for the benefit of the parties and the customer.

Even though the smart home sensorised data generated from a customer's smart home system use is analysed by the insurer to assess current and future risk profiles, the purpose of data collection is very different.

In effect, both the insurer and the smart home system act as data collecting intermediaries for each other to foster mutually beneficial connections for the parties and for the customer. An examination of the Liberty Mutual, State Farm and Canary partnerships is conducted in the discussion that follows to exemplify the operational structure of this model.

4.3.2.1 Operational Data Structure   The Canary partnerships serve as an example of the Partnered Intermediary model in which the smart home device maker, rather than the insurer itself, collects the sensor data. The insurer instead acquires customer data in a post-collection, data-sharing arrangement. For instance, Canary has data-sharing arrangements with a few US-based insurance providers, such as Liberty Mutual, State Farm and Allstate.[56] Other smart home system providers, such as Nest, have similar partnership arrangements with some of the same insurers.[57]

Rather than requiring a Z-Wave hub (as is the case with Neos), the Canary home security system runs exclusively on Wi-Fi. The sensor data is likely sent directly from the Canary device to its cloud-based servers. Likewise, mobile data from the Canary app usage would be sent directly to Canary servers without the need of a specific protocol hub. While there is still a degree of similarity with other partnership models regarding the types of data collected by Canary, there appears to be less emphasis on the amount of sensorised data exchanged with, and ultimately used by, insurers, for customer risk pricing. That said, however, the amount of data exchange is still substantial, and thus 'limited' is construed rather euphemistically by both insurer and smart home system.

The Canary device is also different from the Fibraro home automation system. Canary is a standalone security system that has an inbuilt camera and other sensorised devices, such as a smoke detector and a motion sensor. It is therefore both a hub and several sensorised devices combined.[58] Multiple Canary devices can be used in a single home,

---

[56] The web page for this citation has subsequently been removed. A PDF version is on file with the author. 'How Does Data Sharing with My Insurance Provider Work?' (*Canary*, 2017) <https://help.canary.is/hc/en-us/articles/219571668-How-does-data-sharing-with-my-insurance-provider-work-> accessed 20 June 2019.

[57] 'Insurance Partners' (*Nest*, 2018) <https://nest.com/insurance-partners/> accessed 20 June 2019; 'Customer Agreements for Safety Rewards' (*Nest*, 2018) <https://nest.com/au/legal/customer-agreements-for-safety-rewards/> accessed 20 June 2019.

[58] The web page for this citation has subsequently changed. A PDF version is on file with the author. 'Your Window into Your Home' (*Canary*, 2017) <https://canary.is/how-it-works/> accessed 20 June 2019.

and the manufacturer recommends that a Canary device be installed for each floor.[59] Moreover, the Canary device is exclusive, and other devices cannot be connected through Canary. However, there have been indications that wider integration with other smart home platforms such as Wink and devices such as Amazon Alexa could be taking place.[60] At present, sensorised data collections derive solely from the Canary device itself, but the device is still able to collect a significant amount of data, as outlined in the following discussion.

4.3.2.2 Data Collected Canary collects types of data broadly labelled as 'Personal & Account Information' and 'Device Information' in its privacy policy.[61] 'Limited' data from these broad categories is then shared with insurers under the following headings:

- Canary location data
- Canary owner data
- Canary device settings
- Canary device activity

As noted previously, the Neos privacy policy indicates that all data collected from smart home devices is personal data. However, the Canary privacy policy has a more confined definition of 'personally identifiable information', which appears to exclude device information. The policy states that

> Personally Identifiable Information (PII) is information that contains data that could be used to directly or indirectly identify you, like your name or e-mail address.[62]

PII, Canary states, will only be shared in 'extremely limited circumstances'.[63] This is partly a jurisdictional difference in definitions and policy intentions between the EU (e.g. Neos) and the US (e.g. Canary) that manifests in greater or lesser classifications of sensorised data as personal information and thus covered by relevant information privacy law, as Chapter 7 details in more depth.

---

[59] 'Original Canary FAQ' (*Canary*, 2017) <https://help.canary.is/hc/en-us/articles/206619098-Original-Canary-FAQ> accessed 20 June 2019.
[60] The web page for this citation has subsequently changed. A PDF version is on file with the author. Canary, 'Apple HomeKit and Canary Integrations' (2017) <https://help.canary.is/hc/en-us/articles/221029387-Apple-HomeKit-and-Canary-integrations> accessed 20 June 2019.
[61] 'Canary Privacy Policy, 14 April 2017' (*Canary*, 2017) <https://canary.is/legal/privacy-policy/> accessed 20 June 2019. An updated privacy policy was published in May 2018. A copy of the previous privacy policy is on file with the author.
[62] Ibid.
[63] Ibid.

Like Neos, Canary also collects personal and account information upon set-up of the device.[64] Furthermore, audio and video are collected through the Canary device and stored on Canary servers to allow customers to access playback footage. A substantial amount of sensor data is also collected by the Canary device. In particular, a variety of 'environmental' sensor data is collected, including air quality, humidity, temperature and ambient light levels.[65] Canary's air quality sensor measures ethanol, cooking odours and cigarette smoke in the air, among other things.[66] Device information such as Canary's historic and current configurations, as well as mobile phone information and activity logs, is also collected. Table 4.3 details the type of data collected by the Canary device.

Assuming that the customer opts in for data sharing, Canary then shares a significant amount of data (the 'limited data') with relevant insurance partners – for example, Liberty Mutual, which then provides the customer with a new pricing offer based on analysis of the Canary data. Several of these types of data – such as Canary location, device settings and device activity – are presumably used for the type of risk-based assessments employed in the Local Tapiola/Cozify model. Table 4.4 details the Canary data shared with insurers.

The device data seems particularly pertinent to Liberty Mutual's data analysis, given the purposes of exchanged data proffered by Canary. Again, in some ways this is like Neos' emphasis on customer control data through a customer's use of the Neos app on their phone (hence the collection of gyroscope data, etc. – going back to the coverage of smartphone sensors in Chapter 2).

The Canary terms and conditions also indicate that customer shared data is exchanged to assist the insurer with 'ongoing research, analysis and offerings'.[67] 'Canary information' including 'log data indicating device on/off status, alerts & notifications, sensor readings, and temperature readings' is also shared with the insurer.[68] Customer shared information does not include camera feeds or recordings.[69] All of the information is shared on an opt-in basis, and consent for the sharing can be withdrawn at any time via the Canary website. Canary customer

---

[64] Ibid.
[65] Ibid.
[66] 'What Does Air Quality Measure?' (*Canary*, 2017) <https://help.canary.is/hc/en-us/articles/218911127-What-does-Air-Quality-measure-> accessed 20 June 2019.
[67] 'Save on Your Insurance Premium with Canary' (*Canary*, 2017) <https://canary.is/insurance/> accessed 20 June 2019.
[68] Ibid.
[69] Ibid.

Table 4.3. *Smart home data collected by Canary*

Data Type	c. Details of Data Collected	d. Classed as PII	e. Comment
Personal and account information	Customer information provided during device set-up: name, email address, phone number, the location of each device and the number of people in your home Customer information necessary to complete a purchase: name, billing and shipping address, credit card information and other information necessary to complete your transaction, as well as other customer-volunteered information Other account information: unique identifiers such as usernames or passwords	Yes	
Device information	Audio and visual data Sensorised environmental data: air quality, humidity, temperature and ambient light Device information: model and serial number, Internet Protocol addresses, device activity logs; historic and current device configurations Customer mobile device information: mobile device ID, device type, operating system, service carrier and location (if automatic switching mode is activated)	Yes Unclear/no	Collected in correspondence, with customised modes set by the customer Re automatic switching: 'Some features of our app use location-based data. If you allow these services, we will use information about the Wi-Fi routers and cell IDs of the towers closest to you.'

data sharing is thus optional and does not affect eligibility for an insurance discount, which could be applicable simply through the use of Canary products.[70] Canary states that shared customer data will not affect an insurance premium in a negative way to increase a premium. Finally, if the customer has Canary in multiple locations with different insurance providers, the customer can select a different insurance provider for each location, which again emphasises the intermediary nature of this model.

[70] The web page for this citation has subsequently changed. A PDF version is on file with the author. 'Canary System Terms and Conditions' (*Canary*, 2017) <https://canary.is/legal/system-terms/> accessed 20 June 2019.

Table 4.4. *Canary data shared with insurers*

Data Type	f. Type of Canary Data Collected	g. Data	h. Purpose of Exchange to Insurer
Location data	Personal and account information	Address	Identification of customer insurance policy
	Device data	Auto mode settings; number of location members	Understand how the Canary device is used
Owner data	Personal and account information	Name	Identification of customer insurance policy
	Personal and account information	Account status	Identification of date customer started to use Canary device
Device settings	Device data	Connection state; settings; software version	Understand how the Canary device is used
Device activity	Device data	Sensor readings (temperature, humidity, air quality); alert frequency	Understand how location environment changes over time
			Understand how customer interacts with Canary device through the Canary app

4.3.2.3 Model Characteristics Several characteristic observations about the Partnered Intermediary model can be adduced. The device data collected by Canary is largely similar in type to that collected by Neos. However, given the more expansive scope of Fibraro's system, and the greater number of differently located sensors, it can be assumed that a higher level of sensorised data is collected by the Neos system. That said, collections from both models fulfil largely the same purpose: to be analysed by the insurer to gain a better understanding of customer insights. The respective privacy policies, however, have quite different constructions of personal data and PII. Both models also have quite different opt-in strategies for data collection and exchange. Neos has an expansive definition of personal data that includes sensorised device data with limited opportunities for customer opt-in for certain data collections. Canary has a limited definition of PII which seeks to exclude device data but has more expansive opportunities for customer opt-in regarding the sharing of Canary data with insurers.

The different data collection strategies are influenced by jurisdictional information privacy law compliance, but they are also representative of different underlying data use purposes. The collection/connection element of the Liberty Mutual/Canary model means that there is less emphasis on the requirement to have more detailed knowledge of individual customer behaviours from sensorised collections. The connective component of the Partnered Intermediary model appears to require less sensorised data, thus meaning that 'limited' data can be shared. However, it should be noted that the 'limited' data sharing is still extensive and that several relevant activities or behaviours could be inferred from the data exchanged by Canary.

The focus on connective components also highlights that the foundational roots of the Partnered Intermediary model evolve from different smart home InsurTech developments than those of the Partnered Data Acquisition model. The Partnered Intermediary model evolves from the mutual product promotion and prevention and monitoring services developments, as exemplified by Liberty Mutual's, State Farm's and All State's partnerships with Canary, Nest and other smart home products. The Partnered Data Acquisition model, in contrast, evolves directly from the data acquisition requirement for specific smart home insurance developments. These different evolutionary paths give rise to the different data collection operations, rationales and exchange pathways evident in both models.

For example, the Partnered Data Acquisition model has a predominant focus on substantial sensorised collection in a monogamous and closed relationship between partnership parties. Much of the sensorised data collected is done so automatically, and thus customer opt-in possibilities are reduced. The need for detailed sensorised data to enable the development and delivery of new, customer-focused smart home insurance products and services appears to be the key driver for collection.

The Partnered Intermediary model, by contrast, features a few non-exclusive partnerships between an insurer and smart home systems. An insurer can therefore partner with many smart home systems and vice versa. The partnership relationships are consequently *polygamous and many-to-many* in nature, such that an insurer or a smart home system can partner with a party that has an ongoing partnership with another insurer or smart home system. There is much less emphasis on the collection of smart home sensorised data to facilitate a specific insurance product, which means that there is a greater prominence given to connecting customers with partners, based more heavily on customer opt-in options for data exchange. Essentially, the customer determines the sharing of their smart home data with their dedicated insurance partner in return for new insurance offerings based on an analysis of the shared data.

4.3.3   Platform Entity

The third model is called Platform Entity model, and it is currently more speculative in nature even though it is derived from existing insurer and smart home system partnerships involving Nest. It is speculative because the true power components of this model are still unfolding. However, the advent of home voice control systems by Google and Amazon, as Chapter 3 outlines, is such that it is important to identify the potential parameters of this model now, because it could restrict an insurer's ability to partner with a smart home provider in the future. The same will also apply for any type of business that seeks to utilise smart home data from the Nest suite.

The Platform Entity model shares similar data operation characteristics with the Partnered Intermediary model. On its face, the data exchange mode is also similar, as it appears to feature many-to-many relationships, principally based on promotional discounts and limited exchange of customer data. However, the fundamental difference between the Platform Entity model and the Partnered Intermediary model is evident from its title. An insurer can still partner with many smart home systems, but partnering with a smart home system owned by one the titan platforms, such as Nest owned by Google, may reduce the insurer's ability to set the *conditions* for data collection and partner connection. The platform becomes the data collector and sets the conditions for connection. The insurer is then a partner with a smart home system, but in another sense, it is nothing more than another entity in a titan platform's ecosystem.

The insurer is but a conduit for wider platform service provision that also includes smart home systems. The implicit benefit for the insurer, as with the other models, is that the use of the smart home system encourages customers to respond to potential risk liabilities through greater knowledge of home operations. The increasing popularity of smart home platform products also means that the platform model could increase forms of smart home sensorised data collection as the uptake of products such as Google Home begin to replace specialised smart home hubs such as Fibraro. The vast scope of Amazon's and Google's reach means that the current fragmented infrastructures could be replaced by dominant models of smart home collection regarding the type of voice control system installed.

Moreover, Nest Protect, as a home control system, plays a much greater role in the automated operation of adjusting the home at the expense of customer agency. Unlike in the other models, where customer use of controllers – particularly the smartphone – is encouraged, in the Platform Entity model, the opposite may eventually be the case. While there is the prospect of enhanced forms of home-based sensorised collections, the reality could be the converse, where there is virtually no data collection for the benefit of the insurer. Instead, there is a multiplicity of data collection sources for the benefit of the platform, which then sets the conditions for connections relating to customer collected data.

4.3.3.1 Operational Data Structure   As in the Partnered Intermediary model, in the Platform Entity model, the smart home device system and/or platform, such as Nest or Google, collects the smart home event sensor data from the insured. There is then a similar and subsequent data-sharing arrangement between the platform entity and the insurer. For example, in the United States, partnerships exist between Nest and insurers Liberty Mutual and American Family Insurance. Again, as with Canary, data is collected from a standalone Wi-Fi-protocoled device, Nest Protect, which can be used to connect with a range of other Nest products. Sensor data is sent from the Nest device across Wi-Fi and the Internet, directly to the Nest cloud-based servers. However, while Canary and indeed Fibraro in the Partnered Data Acquisition model have limited capacity for additional device connectivity, Nest Protect has significant capabilities due to its ability to connect with Google Home. While Nest Protect and its limited form of sensorised collection is the current focus of insurer partnerships, the more significant Platform Entity development will be the wide-scale implementation of Google Home. If this happens, then Google and the other titan platforms with their own products place themselves in a position to dominate the smart home market and data collection operations.

A foretaste of the future possibilities is evident, even now. The process of sensorised collection from Nest Protect means that data may be aggregated and/or shared between multiple Nest devices and between Nest servers.[71] For example, Nest's privacy policy notes that if a number of its products are located in a single home, the products will share certain information with each other, such as data on whether something in a room is moving, the temperature and the presence of smoke or carbon monoxide (CO) alarms.[72] Such data aggregation is necessary for the automated adjustment function of Nest products. Aggregation also allows Nest's algorithmic framework to better assess activities in the home, such as automated functions relating to efficient energy use. Again, in terms of Chapter 3, Nest is reflective of Aldrich's attentive home and Lewis' state of intervening. More importantly, the Nest collections are potentially an example of how future data processing might occur on the Platform Entity model, before the data is subsequently shared with insurers or other platform entities. Such data collections indicate that sensorised smart home data collection is not at the direct behest or indirect mutual benefit of insurers; instead, it is all about the platform.

4.3.3.2 Data Collected At present, Nest shares a 'limited' amount of data with insurers. In the United States, customers who are part of Nest's Safety Rewards program are asked to grant Nest permission to provide a monthly status report containing basic summarised information

---

[71] Nest, 'Customer Agreements for Safety Rewards'.
[72] Nest, 'Privacy Statement for Nest Products and Services, 1 November 2017' (2017) <https://nest.com/legal/privacy-statement-for-nest-products-and-services/> accessed 20 June 2019. To Nest's credit, the company has made back-dated versions of its privacy policy available. The research was conducted using the November 2017 privacy policy.

about their Nest Protect to the relevant insurance company.[73] This summarised data includes the status of batteries, smoke sensor, carbon monoxide sensor and connection to the Internet. The status report is limited to basic values such as:

- 'Good' – functioning normally
- 'Low' – battery charge is low
- 'Issue' – problem with one or more sensors
- 'Unknown' – there may be an issue, but Nest Protect cannot diagnose it or has not checked in because it is offline[74]

The status report also includes the customer's postcode and the names of the rooms in which Nest Protect is installed.[75] In receiving this information, the insurer can verify that the Nest Protect products are working properly and are affording the insurance customers the safety benefits of Nest products. Sensorised data thus far appears to only be collected from the Nest Protect device for insurance exchange purposes. Although it is possible for a customer to purchase other Nest devices such as the Nest thermostat or camera, which can then be connected to Nest Protect, it is not clear whether the additional device data is also exchanged with insurance partners.

Nest's privacy policy states that the following types of data are collected through Nest Protect:

- set-up information a customer provides
- environmental data from the Nest Protect's sensors
- technical information from the device
- additional information

These categories of data are broadly similar to those of both the Neos and Canary models. Interestingly, Nest has the longest privacy policy when compared to those of Neos and Canary. However, the policy does not define 'personally identifying information', 'personal information' or 'personal data'. Personal information in the three policies is either defined expansively (e.g. Neos) or narrowly (e.g. Canary) or not defined at all (e.g. Nest). Chapter 7 discusses these implications in greater depth. Table 4.5 details the data collected from Nest Protect.

---

[73] 'When I Enroll in Safety Rewards, What Kind of Data Is Shared with My Insurance Company?' (*Nest*, 2017) <https://nest.com/support/article/When-I-enroll-in-Safety-Rewards-what-kind-of-data-is-shared-with-my-insurance-company> accessed 20 June 2019.
[74] Ibid.
[75] Ibid.

Table 4.5. *Smart home data collected by Nest Protect*

i. Data Type	j. Details of Data Collected	k. Purpose of Collection
Set-up information	Information such as home address or postcode, location of sensor installation in the home, username, email, address, etc.	Enable customised experience based on analytics; check ongoing operation of sensorised devices
Environmental data	Sensor-generated collections related to smoke and CO levels, current temperature, humidity, ambient light and motion detection	Enable Nest Protect to detect risks and provide warnings
Technical information	Nest Protect model and serial number, software version and technical information such as sensor status, Wi-Fi connectivity and battery charge level	Improve customer experience and troubleshooting
Additional information: Wi-Fi network information	Collected during set-up: Wi-Fi network name (SSID) and password to connect to the Internet; IP address	Enable connection of Nest Protect to broader services via Wi-Fi network
Additional information: Additional authorised users	Information about invited users (e.g. email address, name, changes to product settings)	Enable addition of authorised user to customer account
Additional information: email addresses	Customer email address	Communication; enable authorised users
Additional information: Basic profile information	Name, photo, authorised username and photo	Enhance public profile
Additional information: Mobile location data	Location data from mobile device	Enable location-based features and device pairing
Additional information: Bluetooth data	Not stated	Connect to other Bluetooth devices

Like the other models, a significant amount of data is collected through Nest Protect, some of which is shared with insurance partners, Liberty Mutual and American Family, for promotional discounts and product offers. It is also evident that a significant amount of data aggregation is conducted with other Nest services. This perhaps gives an insight into the future data exchange possibilities arising from this model. Currently, Nest aggregates sensorised data from multiple

Nest devices in one home. Given Google's ownership of Nest, it could become possible to also aggregate Nest data with other data from Google Home and other Google services, such as Google Assistant and Google Search. While Nest currently operates from the perspective of an intermediary, there is a realistic prospect that it will evolve further into Google's broader platform ecosystem.

4.3.3.3 Model Characteristics   The following observations are put forward. As noted earlier with respect to Neos and Canary, the exact process of data collection in the Platform Entity model will depend on the devices and protocols that the relevant customer decides to use. However, of the three models, the Platform Entity model will probably allow for the largest amount of data to be collected, as smart home platforms such as Google's and Amazon's will be interoperable with a wider range of smart home services and devices. Thus, the smart home platform will be able to collect data not only from a customer's usage of a single device – for example, Nest Protect – but also of all other third-party devices in that home. With a larger volume of data potentially available in this model, more granular and accurate insights would be possible for data analytics, as opposed to partnering with single smart home device makers in an ad hoc fashion. Consequently, this model may offer more data analysis opportunities in terms of the richness and volume of data, but it also gives rise to the prospect of less insurer control over collection and connection strategies.

It is also uncertain whether the platform would act merely as an intermediary that simply passes on raw sensor data to its data-consuming partners, such as insurers. It is more likely that a platform will undertake a degree of data processing before sharing the smart home data with third parties. In this regard, the platform is likely to fulfil a *condition-creation* role in setting platform rules on how to aggregate or process smart home data before sharing with insurers. As mentioned, the insurer, or any other business, as a platform entity, may have less control over data collection and processing decisions as compared to the previous two models, the Partnered Data Acquisition and Partnered Intermediary models.

On its face, the Platform Entity model has many similarities to the Partnered Intermediary model in seemingly involving polygamous, many-to-many partnership relationships. However, on closer inspection, this evolving model is potentially quite different and could thus be categorised as a *dominant form of polygamous relation* that actually features a one-to-many relationship in which the 'one', the titan platform, is dominant in terms of market share of implemented smart home products and thus creates the conditions for data collection and the subsequent data exchange connections with other platform entities.

4.4   Conclusion

The smart home insurance models highlighted in this chapter are important to understand as part of the book's overall argument. The models demonstrate that the use of sensorised data is becoming a key constituent for business model development based on partnership arrangements. Smart home insurer and smart home system partnerships have existed since 2013. The first partnerships did not involve the exchange of sensorised data collections, but that is now increasingly changing. The evolutionary development of smart home insurance partnerships is evident in the increasing complexity of partnership models.

Smart home automation or control systems, based on the component infrastructure outlined in Chapter 3, appear to be significant partners of interest for established insurance companies seeking to expand their technological ambits in relation to home insurance services or products. Such partnerships are being beneficial to both entities, as insurance companies have the capacity to deal with the more capital- and regulatory-intensive aspects of the insurance sector, while smart home start-ups have the technological know-how, particularly in relation to sensorised data collection technologies.

The foundational basis for these partnership arrangements is the sharing of new or existing datasets about customers. Established insurers have data accumulation advantages both in terms of quantity and longevity. In other words, they have more customers and more historical data about those customers, including about their homes. Home automation systems and start-ups, on the other hand, have the technical expertise, products and capacity to generate new forms of sensorised collections that are crucial to understanding the day-to-day behaviours of individual customers, captured through sensor data. As highlighted in Chapter 3, it is this sensorised data that provides the necessary data inputs to enable real-time analysis that could power behavioural nudging mechanisms.

It is possible to identify key differences in rationale between these models that indicate that different business models around sensorised data collections are in operation. All these models offer control challenges for commercial entities seeking to partner in the smart home space.

The Partnered Data Acquisition model offers complete control over collected sensorised smart home data. However, as the smart home continues to develop, this model's ability to collect smart home data may diminish if consumers continue to favour other smart home products. The Partnered Intermediary model offers less control over collected data but offers more opportunities to partner with multiple smart home systems, thus obviating the risk of diminishing obscurity arising from the Partnered Data Acquisition model. The construct of control is replaced with mutuality, because this business model requires a greater focus on connection with mutual beneficiary partners rather than control of collected data. The final model, Platform Entity, potentially divests a sole data collector's ability to control either data collection or connection. Those aspects are built into the condition-creation aspects of dominant 'platformatisation'.

Notions of control regarding sensorised data collections from the home, and indeed from the broader collected world, are clearly important. The smart world is offering new forms of data collection for private and public sector entities that are increasingly aimed at identifying the patterns and behaviours of individuals or broader populations. As such, much, if not all, of the sensorised data collected could be classified as personal information and thus be regulated by information privacy law.

These new sensorised opportunities will give rise to some significant challenges for all parties in the data collection process. The manifest form of sensorised data collections that are currently unfolding diminishes the ability of data providers, collectors and secondary users to exhibit control over collected data. An evolutionary tension thus unfolds across technological and legal spectrums: sensorised environments enhance data collection possibilities but provide less ability to control data acquisition and use. The tension gives rise to a knock-on effect for information privacy law which is imbued on notions of control. It is unclear how the law applies in relation to control-diminished environments, which feature ever-increasing opportunities and abilities to collect sensorised data for individual monitoring purposes. A key question therefore arises – *to what extent can information privacy law continue to provide appropriate legal protections for individuals when data about everything is collected?*

Now that the frame of the collected world has been established, Parts II and III of the book examine the legal consequences that will arise to address this challenging question. Part II provides a conceptual and application overview of information privacy law. Part III then examines the application of some key elements of information privacy in the face of the collected challenges outlined in Part I. The book then concludes with suggested responses in relation to the conceptual and practical application of information privacy law.

Before we get that far, however, it is necessary to ask two seemingly simple questions. What does information privacy law seek to protect? How does information privacy law protect? As we will see, the answers to these deceptively simple questions are relatively complex.

Part II

Information Privacy Law's Concepts and Application

5   What Information Privacy Protects

5.1   Introduction

Part I of the book outlined a world that is becoming increasingly connected and the collected consequences that flow from ubiquitous
connectivity. Chapters 3 and 4 gave hints to some of the information
privacy law consequences that are likely to emerge. The second part of
the book delves further into information privacy law and explores its
conceptual basis and substantive processes of application across two
key components. Chapters 5 and 6 establish the legal framework for
examining the concerns arising from ubiquitous collections, such as
those redolent in the smart home, as a foundation for outlining the collected challenges covered in Part III.

The purpose of this chapter is to provide a conceptual overview of
information privacy law, to examine what it seeks to protect. This is
not a straightforward task,[1] given that both privacy and information
privacy are essentially contested concepts.[2] Different concepts of information privacy provide alternative ideas about what its legal manifestations should seek to protect. Moreover, information privacy concepts are
themselves permeated, and permeate, broader concepts of privacy. How
information privacy manifests – namely, as a right, a value, a state or a
claim – also corresponds to the intrinsic questions found in the broader
privacy law literature.

Nonetheless, despite these differences, there are also some core
commonalities found in information privacy concepts, particularly
regarding the role of information exchange. Information privacy, at
its core, seeks to protect information intrinsic to human beings and

---

[1] Cohen, 'Turning Privacy Inside Out', 3 stating that concepts of privacy are built on
   contradictions.
[2] Deirdre K Mulligan, Colin Koopman and Nick Doty, 'Privacy Is an Essentially
   Contested Concept: A Multi-Dimensional Analytic for Mapping Privacy' (2016)
   374 Philosophical Transactions of the Royal Society A 1.

the lives they live. As demonstrated in Chapter 6, the legal expression of this type of information typically regards jurisdictional classifications of personal information, personal data or personally identifying information. Information privacy law therefore provides individually focused protections for specified categories of information.

Information privacy, on its face, reduces the richness of broader concepts of privacy. However, that does not mean that information privacy lacks conceptual depth. Far from it. Information privacy seeks to imbue values and protect expectations relating to the essential social practices of information exchange that are intrinsic to the functioning of our everyday lives.

In this chapter, we examine four conceptual themes that house a range of different perspectives. Before we do, however, it is important to note that the study of privacy law is rich in depth, complexity and intellectual rigour. The concept of privacy is an essentially contested subject, and many different conflicts exist regarding privacy's conceptual reach and legal application.[3] Much academic ink has been spilt on attempts to conceptually define privacy from a legal perspective. Attempts to answer the question 'What is privacy?' in a meaningfully legal sense have generated a literature that is immense in its intellectual breadth, intense in its scholarly conviction and ingenious in its development of analytical frameworks. However, an answer to the question sought has not been forthcoming, thus leading to a degree of despair and confusion about whether a straightforward solution can ever be found. To a lesser extent, the same argument arises in relation to information privacy law, as exemplified through the coverage of four key conceptual themes.

The themes have been parsed to provide a coherent, and relatively chronological, concept map of information privacy's broader development. However, as will be evident in the oncoming coverage, the intersection of different conceptual perspectives is complex and often overlaps the critiques and concepts of different authors. For example, the coverage of Alan Westin is predominantly focused on the genesis of control-oriented concepts, but his work provides telling insight into broader autonomy considerations. Similarly, Stanley Benn's work is covered predominantly in its autonomy context but has strong echoes of control-based concepts. The relational coverage of Helen Nissenbaum and Julie Cohen also derives concerns about power, as do Benn's considerations of heterarchy.

---

[3] Daniel J Solove, 'Conceptualizing Privacy' (2002) 90 California Law Review 1087; Anita Allen, 'Synthesis and Satisfaction: How Philosophy Scholarship Matters' (2019) 20 Theoretical Inquiries in Law 343.

Nevertheless, the thematic classification adopted in this chapter is assistive regarding the consideration of collected challenges, consequences and responses found in Part III of this book, particularly in relation to the power context and the overlaps with other concepts.

The conceptual themes are selected specifically with regard to the information privacy context and thus focus predominantly on implications for information exchange. Koops et al. have argued that informational privacy occupies a centrifugal space around which other privacy issues and concerns gravitate.[4] Different aspects of privacy law, such as communications privacy and associational privacy, also have significant informational components. Their work is helpful because it connects the role of information privacy with broader concepts of privacy, such as secrecy[5] or intimacy.[6] The conceptual focus of the book, however, is specifically on the role of information privacy law in the face of increasingly ubiquitous collections of sensorised data. This chapter consequently focuses on the conceptual and practical role of the law regarding processes of information exchange. The four themes adduced are representative of this narrower approach.

The first theme focuses on the importance of an individual having control over their information, including who can access their information. Information privacy in this regard is an individually focused protection that seeks to safeguard user control over specified categories of information exchange. Control theories of information privacy often equate questions of control with questions of ownership. In doing so, control theories tell us what information needs to be protected and how it could be protected, but there is an often under-articulated subtext as to what control protections ultimately provide regarding exchange decisions.

The underlying subtext behind control is perhaps articulated more clearly in the second theme: information privacy as a protection of personal autonomy. The historical focus on this theme finds its roots in control theories of information privacy but seeks to broaden the considerations beyond individual control mechanisms. A relational context is thus tacitly introduced, as information privacy is conceptually considered an essential societal component of liberal societies and a concomitant requirement for the flourishing of autonomous individuals.[7] The second theme therefore provides a broader framework to critique the notion that

---

[4] Bert-Jaap Koops and others, 'A Typology of Privacy' (2017) 38 University of Pennsylvania Journal of International Law 483.
[5] Richard A Posner, 'The Right of Privacy' (1978) 12 Georgia Law Review 393.
[6] Jeffrey Rosen, *The Unwanted Gaze: The Destruction of Privacy in America* (Random House 2000).
[7] Cohen, 'What Privacy Is For' 1904, 1905.

information privacy should only provide individually focused protections. In other words, information privacy as a protection of autonomy automatically has a wider societal and relational component.

Information privacy in the third theme, the relational context, critically questions the conceptual bases of individual control mechanisms that protect informational spaces for autonomous growth. The third theme examines more broadly the societal and relational consequences of information exchange. Information privacy, in this sense, regards individual protections as an intrinsic societal fabric that is interwoven with different social practices and relationships involving information exchange. The third theme thus critiques the basis of information privacy as a sole protector of individual control mechanisms and casts doubt about the focus on individual informational spaces that enable personal autonomous growth.

The growth of control-based critique leads us to the final theme and perhaps the most controversial concept of information privacy, as a problem of power relations. The final theme questions the structural foundations of relational information exchange and critically examines who constructs and benefits from these societal structures. These arguments are most recently expressed as surveillance capitalism, as outlined briefly in the discussion that follows.

5.2   Individual Control over Personal Information

As alluded to in Section 5.1, concepts and notions of information privacy are intersected with other differing, and sometimes competing, concepts of privacy.[8] Both the concepts of privacy and information privacy have consequently been difficult to define from a legal perspective. Take, for example, this small snapshot of different viewpoints.

The dean of Yale Law School, Robert Post, has commented that the notion of privacy is so complex that it cannot be usefully conceptualised because it is so entangled with competing and contradictory dimensions.[9] Anita Allen, who adopts a feminist perspective to privacy, sees privacy as an inalienable right that should be considered as a preconditional foundation of a liberal egalitarian society.[10] Privacy has also been

---

[8] Tamara Dinev and others, 'Information Privacy and Correlates: An Empirical Attempt to Bridge and Distinguish Privacy-Related Concepts' (2013) 22 European Journal of Information Systems 295.

[9] Robert C Post, 'Three Concepts of Privacy' (2001) 89 Georgetown Law Journal 2087.

[10] Anita L Allen, 'Coercing Privacy' (1999) 40 William and Mary Law Review 723, 745.

conceptualised as 'concern for limited accessibility', which provide barriers and limits regarding the extent of what is known about us and physical access to us.[11] Julie Inness has highlighted the importance that privacy has regarding the intimacy of the individual.[12] Paul Schwartz further contends that privacy is integral, as it forms the basis of experimental development of intimate thoughts, and that anonymity is a key determiner of the maturity of personhood, especially in a world increasingly dominated by instant communication and access.[13]

Already we can see that several different yet overlapping concepts emerge; privacy as a foundation for liberal society, privacy as a protection of access to ourselves, privacy as a space that allows us to develop and mature. Attempts to conceptually define privacy have thus been challenging because of the multifaceted nature of the concept. Despite the differing ideas, the legal development of information privacy is most generally associated with control concepts of privacy that seek to protect individual control in decisions about the use of personal information.[14]

The control concept is foundationally important because it has become the dominant paradigm of information privacy law. As outlined in Chapter 6, the basis of information privacy law is predicated upon providing control mechanisms to individuals to ensure fairness in information exchange processes. However, the effect of the policy-making process has contracted the breadth of control concepts in law to the extent that legal protections in some jurisdictions have largely become procedural.[15] The overt focus on procedural protections relating to personal information has left control concepts largely detached from the broader conceptual base from which they were initially formed.[16] The control roots of information privacy are broad, tangled and deep, but their visible legal flourishment has not necessarily reflected their conceptual complexity.[17]

---

[11] Gavison.

[12] Julie C Inness, *Privacy, Intimacy, and Isolation* (Oxford University Press 1992).

[13] Paul M Schwartz, 'Privacy and Democracy in Cyberspace' (1999) 52 Vanderbilt Law Review 1609.

[14] Colin J Bennett, *Regulating Privacy: Data Protection and Public Policy in Europe and the United States* (Cornell University Press 1992) 14 regarding the analogous links between 'data protection' and Westin's information privacy; Lisa Austin, 'Privacy and the Question of Technology' (2003) 22 Law and Philosophy 119, 125; Herman T Tavani, 'Philosophical Theories of Privacy: Implications for an Adequate Online Privacy Policy' (2007) 38 Metaphilosophy 1, 7.

[15] Michael Birnhack, 'A Process-Based Approach to Informational Privacy and the Case of Big Medical Data' (2019) 20 Theoretical Inquiries in Law 257, 260.

[16] Ibid. Note, however, that Birnhack lined privacy with the 'concretization of human dignity, translated into privacy as control.'

[17] Valerie Steeves, 'Theorizing Privacy in a Liberal Democracy: Canadian Jurisprudence, Anti-Terrorism, and Social Memory after 9/11' (2019) 20 Theoretical Inquiries in Law 323, 327.

The genesis of control concepts in the 1960s is heavily linked to the advent of newly computerised systems in both public and private sectors. The early progenitors of control concepts, aware of the vast structural changes taking place, highlighted the importance of individual control over information as the basis for future legal protections of privacy. Charles Fried, for example, argued that '[p]rivacy...is the *control* we have over information about ourselves'.[18] Control over information was intrinsically linked to personal liberty and as a basis for the establishment of loving and intimate relations.[19] Arthur Miller, in a wide-ranging examination of the key privacy/invasive technological developments of the period also defined privacy in the same capacity – as 'the individual's ability to control the flow of information concerning or describing him'.[20] Miller similarly outlined the importance of the control notion of privacy as a basis for establishing close relationships and for the maintenance of personal freedoms. The advent of new technological frameworks – most notably the ability to collect a wider range of information and store it in increasingly centralised databases – could compromise an individual's ability to control flows of information about them.[21] The ability of individuals to control information was subsequently important. Without control protections, individuals would be more likely to make personal decisions in line with the expectations of third-party data collectors who would thus control important aspects of their lives.[22]

The substantive link between broad data collections and the power to shape individual activities and decisions was thus made relatively early in the newly computerised environment. As already evident, the early notions of control-based informational privacy had much broader notions of the purpose of individual control than was to subsequently manifest in the early forms of information privacy law. The same can also be said[23] of the most influential representation of the control concept, Westin's *Privacy and Freedom*.[24] Interestingly, unlike Fried and Miller mentioned earlier, Westin did not use either the term 'right' or 'control' or even 'information privacy' in his description of an

---

[18] Charles Fried, 'Privacy' (1968) 77 Yale Law Journal 475, 482.
[19] Ibid 481–3.
[20] Arthur R Miller, 'Personal Privacy in the Computer Age: The Challenge of a New Technology in an Information-Oriented Society' (1968) 67 Michigan Law Review 1091, 1108.
[21] Ibid 1109.
[22] Ibid 1125. Note here the crossover with autonomy and power considerations highlighted earlier.
[23] Austin, 'Re-reading Westin'.
[24] Alan F Westin, *Privacy and Freedom* (Atheneum 1967).

individual's required claim for informational privacy.[25] Nevertheless, his work has been perceived as the clarion call for the provision of individual rights of control over personal information.[26]

In *Privacy and Freedom*, Westin determined four basic states of individual privacy:[27] *solitude, intimacy, anonymity* and *reserve*.[28] The latter state, reserve, is of most interest regarding control concepts of information privacy. The state of reserve requires the

creation of a psychological barrier against unwanted intrusion; this occurs when the individual's need to limit communication about himself is protected by the willing discretion of those surrounding him.[29]

Westin argued that the need for individuals to have barriers is necessary, as the communication of the self is always incomplete. Individuals are required through their ongoing involvement in society to retain some information about them which is too personal for other persons or organisations to possess.[30] This mental distance – the space generated by choosing not to declare everything about one's self – therefore requires an individual to have the ability and control to withhold or to disclose personal information. The ability of choice over our own information is consequently the 'dynamic aspect of privacy in daily interpersonal relations'.[31]

Westin also adduced four specific functions of privacy that reflect the value or purpose of privacy within society. They are *personal autonomy, emotional release, self-evaluation* and *limited and protected communication*. Again, the latter function is of relevance to control theories of privacy,

---

[25] Note, however, ibid 7 regarding Westin's 'right of individual privacy', which is defined as 'the right of the individual to decide for himself, with only extraordinary exceptions in the interests of society, when and on what terms his acts should be revealed to the general public'.

[26] Raymond Wacks, *Personal Information: Privacy and the Law* (Clarendon Press 1993) 14 noting the influence of privacy and freedom in relation to understandings of privacy as control of personal information; James B Rule, *Privacy in Peril* (Oxford University Press 2007) 22 regarding the influence of Westin's work and the need to regulate organisational data systems in the late 1960s and early 1970s; James Waldo, Herbert Lin and Lynette I Millett, *Engaging Privacy and Information Technology in a Digital Age* (Information Privacy, National Academies Press 2007) 60 highlighting Westin's role in the development of the concept of information privacy.

[27] Austin, 'Re-reading Westin' 60: all four states involve social withdrawal, but all states entail information.

[28] Westin 31–2.

[29] Ibid 32.

[30] Ibid.

[31] Ibid.

and it has two facets. The first, limited communication, sets interpersonal boundaries for the exchange of personal information. The second, protected communication, 'provides for sharing personal information with trusted others'.[32] It is the state of reserve in conjunction with limited and protected communication that is inherent in Westin's classic definition of information privacy:

Privacy is the claim of individuals, groups, or institutions to determine for themselves when, how, and to what extent information about them is communicated to others.[33]

As we will see in Chapter 6, Westin's work has been vastly influential, to the extent that information privacy law is founded on the idea that individuals have rights relating to control over their personal information,[34] or at least have rights pertaining to who can access their personal information[35] or a combination of both.[36]

However, the 'privacy as control paradigm',[37] which is most associated with Westin's work, is not without its critics. Schwartz highlighted that whilst the control model has benefits because it seeks 'to place the individual at the centre of decision-making about personal information use',[38] it nonetheless suffers from several major flaws because it pays little consideration to information asymmetries – a point that will be

---

[32] Stephen T Margulis, 'On the Status and Contribution of Westin's and Altman's Theories of Privacy' (2003) 59 Journal of Social Issues 411.

[33] Westin 7.

[34] Miller 1107 'the basic attribute of an effective right to privacy is the individual's ability to control the flow of information concerning or describing him'. See also Priscilla M Regan, *Legislating Privacy: Technology, Social Values, and Public Policy* (University of North Carolina Press 1995) 9 commenting that privacy, in regard to US governmental collection of personal data, was defined as the 'right of individuals to exercise some control over the use of information about themselves'; Jerry Kang, 'Information Privacy in Cyberspace Transactions' (1998) 50 Stanford Law Review 1193, 1203 referring to an individual's control over the processing of personal information.

[35] Gavison 423 contending that privacy is a concern of accessibility that includes physical access by and the attention of other individuals; Rule 3 'let me define privacy as the exercise of an authentic option to withhold information on one's self'; Solove 1110 stating that information privacy as the right to 'control-over-information can be viewed as a subset of the limited access conception'.

[36] James H Moor, 'Towards a Theory of Privacy in the Information Age' (1997) 27 Computers and Society 27; Adam Moore, 'Privacy, Speech and the Law' (2013) 22 Journal of Information Ethics 21.

[37] Paul M Schwartz, 'Internet Privacy and the State' (2000) 32 Connecticut Law Review 815, 820.

[38] Ibid 822.

covered throughout but particularly in the fourth conceptual theme.[39] Daniel Solove also questions the basis of control as 'privacy self-management' and its ability to foster processes of meaningful control for individuals.[40] Regan also states that Westin's work is applied from an individualistic perspective, which leads to the conclusion that Westin regarded 'privacy as fundamentally at odds with social interests'[41] when that is clearly not the case.[42] Similarly, Hartzog argues that the notion of individual control is far too precious and finite to meaningfully scale.[43] Moreover, criticism is levelled at privacy as control, from the seemingly tautological perspective that privacy as control is either too broad or too narrow.[44]

Allen also contends that there is a fundamental disconnect between what can be considered as having control over personal information and the requirements of a sufficient state of privacy, because the former is not necessarily a constituent element of the latter.[45] Instead, information privacy as control directs attention to issues of consent and choice about uses of personal information that connote an element of inaccessibility separate from privacy considerations.[46] The control aspect of information privacy has also been subject to criticism.[47] Simitis contends that privacy considerations no longer arise out of individual problems but that they instead express conflicts that affect everyone. Information privacy is consequently not simply a problem of individual control over information, as outlined in the remaining themes.[48]

[39] Ibid 830 regarding privacy as control as the 'commodification illusion'.

[40] Daniel J Solove, 'Privacy Self-Management and the Consent Dilemma' (2013) 126 Harvard Law Review 1880.

[41] Regan, Legislating Privacy 28. See also Colin J Bennett and Charles D Raab, The Governance of Privacy: Policy Instruments in Global Perspective (MIT Press 2006) contending that Westin undertook a functional view regarding his investigation of privacy for an individual.

[42] Regan, Legislating Privacy 220.

[43] Hartzog 426.

[44] Solove, 'Conceptualizing Privacy' 1112 contending that privacy as control is too vague due to the failure to define the types of information that individuals should control, whilst other theories overcompensate and becoming too limiting.

[45] Anita L Allen, 'Privacy as Data Control: Conceptual, Practical and Moral Limits of the Paradigm' (2000) 32 Connecticut Law Review 861, 867–8 regarding the differences between physical and informational privacy.

[46] Ibid 869 stating that informational privacy involves information in a state of inaccessibility.

[47] Austin, 'Privacy and the Question of Technology' 125.

[48] Spiros Simitis, 'Reviewing Privacy in an Information Society' (1987) 135 University of Pennsylvania Law Review 707, 709.

Another key element of Westin's work that has been subject to much criticism is the equation of information privacy with property ownership. Westin states:

[P]ersonal information, thought of as the right of decision over one's private personality, should be defined as a property right, with all the restraints on interference by public or private authorities and due-process guarantees that our law of property has been so skilful in devising.[49]

Westin's notion of personal information as property has clear echoes with Warren and Brandeis' seminal consideration of fledgling privacy protections at the end of the nineteenth century based on inviolate personality.[50] Warren and Brandeis were early-day protagonists of media restraint. Gossip, which was once confined to a discrete number of individuals, was increasingly becoming a new industry. They concluded that a demonstrable difficulty arose with current legal protections because the type of harm that arises from privacy invasions invariably involve emotional or reputational harm – a harm that the law, by and large, did not recognise as being sufficiently commensurable. As such, Warren and Brandeis considered that existing protections, such as intellectual property protections, were in fact a mere 'instance of the enforcement of the more general *right to be let alone*'.

The principle which protects personal writings and all other personal productions, not against theft and physical appropriation, but against publication in any form, is not the principle of private property, but that of an *inviolate personality*.[51]

A new legal remedy was therefore required to protect the privacy of individuals, particularly from the media and other persons who possessed new technologies that had the capacity to intrude upon an individual's privacy. Accordingly, Warren and Brandeis' 'right to be let alone' is a reflection of a wider concept – namely, the right to protect one's personality and the right to an inviolate personality is in itself based upon a broad conception of property rights.[52] However, the right to privacy as defined was not an absolute right, and the privacy requirements of individuals must 'yield to the demands of public welfare or of private justice', whilst acknowledging that defining these boundaries would be no easy task. As such, a new law was required to

---

[49] Westin 324–5.
[50] Samuel D Warren and Louis D Brandeis, 'The Right to Privacy' (1890) 4 Harvard Law Review 193.
[51] Ibid 205.
[52] Ibid. See Birnhack 263 linking 'inviolate personality' and control regarding broader dignity theories.

[p]rotect those persons with whose affairs the community has no legitimate concern, from being dragged into an undesirable and undesired publicity and to protect all persons, whatsoever; their position or station, from having matters which they may properly prefer to keep private, made public against their will.[53]

The brief coverage of Warren and Brandeis will become relevant in Chapter 6. The notion of personal information as property and control concepts of information privacy are an important constituent of a broader tradable framework of data that is indicative of the foundational differences between jurisdictions – most notably, the United States and the EU. The conceptual roots of control and concomitant notions of property allocation are more deeply engrained and entangled in the United States than in other jurisdictions, which has influenced how information privacy laws have developed. It should also be no surprise, given the impact of Westin's and Warren and Brandeis' work, that subsequent, and prominent, US law academics have attempted to develop this idea further.[54]

Notions of control-based information privacy protection thus morph into considerations of economic transaction.[55] In these guises, control concepts regard information provision by an individual that is based on what economic interest can be derived from the supply of personal information to a data-collecting organisation.[56] Information, as highlighted previously, is supplied within the context of protecting the individual's most prized asset – their reputation.[57] The decision about disclosing personal information thus becomes a cost-benefit analysis which is decided by balancing the impact of disclosure against the damage to reputation. Consequently, 'an initial model of informational privacy could be to permit disclosure if and only if:

1. The value of the information when disclosed exceeds the value of the pure privacy preference of the individual; and
2. Permitting disclosure will not distort or eliminate the information in future transactions.'[58]

---

[53] Warren and Brandeis 214.
[54] Richard S Murphy, 'Property Rights in Personal Information: An Economic Defense of Privacy' (1996) 84 Georgetown Law Journal 2381; Lawrence Lessig, 'Privacy as Property' (2002) 69 Social Research 247.
[55] Austin, 'Re-reading Westin' 69 regarding property arguments: 'property functions as a kind of proxy for privacy'.
[56] Nadezhda Purtova, 'The Illusion of Personal Data as No One's Property' (2015) 7 Law, Innovation and Technology 83, 102.
[57] Murphy 2385.
[58] Ibid 2387.

Privacy rules are therefore viewed by Murphy as implied contractual terms. Reuse issues are governed by the means through which personal data is collected and the value that is assigned to a person's personal information. The notion of privacy as property ownership characterises personal information as a wealth that can be stolen.[59] Individuals should therefore be given effective control instruments to protect their information assets, and in doing so, protect themselves.[60]

As noted earlier, the idea of control concepts of information privacy as property transaction is very much grounded in the American literature.[61] Not surprisingly, it has also been a subject of debate and criticism. While it seems to have relevance at face value, Julie Cohen argues that a closer inspection of underlying concepts reveals that it has limited application as a resolver of information privacy problems because the concept focuses on a single, notional characterisation of the problems.[62] The protection of intellectual property has also shown that the overt use of piracy surveillance by companies seeking to protect copyright 'has inverted the relationship between privacy and property, subordinating the protection of privacy to the protection of property'.[63] Information privacy as property transaction therefore unduly emphasises the notion of a market-based approach to information privacy, and personal information can be subject to market exchange.[64] These views place the economic interests of a free market information economy above the requirements of the individual,[65] which has become a common refrain of the US information privacy regime, discussed in Chapter 6.

As previously noted, control notions of informational privacy are at the heart of many different ideas of privacy. Many privacy problems can be categorised, in some way, as a loss of control over personal information

---

[59]   Adam D Moore, 'Intangible Property: Privacy, Power and Control' in Adam D Moore (ed), *Information Ethics: Privacy, Property, and Power* (University of Washington Press 2005). 'Stolen' in Moore's sense means unwarranted access or lack of control over information exchange.

[60]   Corien Prins, 'Property and Privacy: European Perspectives and the Commodification of Our Identity' in Lucie M C R Guibault and P B Hugenholtz (eds), *The Future of the Public Domain* (Information Privacy, Kluwer Law International 2006) 223; Moore 24.

[61]   Jessica Litman, 'Information Privacy/Information Property' (2000) 52 Stanford Law Review 1283.

[62]   Julie E Cohen, 'Examined Lives: Informational Privacy and the Subject as Object' (2000) 52 Stanford Law Review 1373, 1391.

[63]   Sonia Katyal, 'Privacy vs. Piracy' (2004) 7 Yale Journal of Law & Technology 222, 228.

[64]   Cohen, 'Examined Lives' 1381.

[65]   Alan Charles Raul, *Privacy and the Digital State: Balancing Public Information and Personal Privacy* (Kluwer 2002).

and the unauthorised disclosures of that information. Information privacy thus has a broader role to play regarding privacy protections and also a broader role regarding the application of other areas of law. The notion of control is therefore important. If an individual can control their personal information, then they can shape how decisions affecting them are made. They can freely establish intimate relationships. The ability to control provides individuals the freedom to think what they want to think and act without impediment. Informational control thus intrinsically serves the broader purposes of autonomous growth that are required for liberal societies to function, a point that has perhaps been diminished with an increased focus on control from a property allocation and transaction perspective. Hence the importance of the second theme – information privacy as a protection of personal autonomy through informational space for personal autonomous growth.

5.3   Informational Access and Personal Autonomous Growth

One of the defining characteristics of control concepts of information privacy regards the clear delineation of information that could be classified as private and public.[66] As a consequence, actions that regard protections in privacy law, or information privacy law, have traditionally been linked to information that is capable of being classified as private. Solove has argued that this approach is problematic, as different concepts of privacy, including control concepts of information privacy, that have been generated on the basis of a public-versus-private dichotomy are either too broad or too narrow.[67] The broadness emanates because some concepts fail to exclude situations that are not related to private actions; the narrowness arises because some concepts fail to include aspects that some would consider private.

The common-denominator approach consequently has an inherent difficulty at its heart – the ultimate choice of denominator. If the denominator is too broad, that gives rise to a concept of privacy that is too vague, as a multitude of different problems could be defined as privacy problems. If the denominator is too narrow, that could give rise to a concept that is too restrictive in application and would not cover a range of problems that one would generally expect to be privacy

---

[66] Judith Wagner DeCew, 'The Conceptual Coherence of Privacy as Developed in Law' in Ann E Cudd and Mark C Navin (eds), *Core Concepts and Contemporary Issues in Privacy* (Springer 2018).
[67] Solove, 'Conceptualizing Privacy' 1094.

problems.[68] Information privacy is often singled out as an example of the latter complaint, as it reduces the richness and scope of privacy to matters involving the control over personal information.[69]

The public-versus-private dichotomy as a means of conceptually defining privacy has therefore been subject to some criticism. Nissenbaum argues that the dichotomy is used as a basis for clarifying a reduced scope of privacy law protections that entail private actors, in private domains that involve private information.[70] This conception greatly reduces situations that could otherwise be considered privacy problems, but it also ignores the complexities of contemporary technology usage by aligning privacy strictly within a private sphere. As such, anything goes in the public domain, which does not sufficiently account for technological developments such as enhanced device sensorisation or the smart home, in which the boundaries of what is public and private have become increasingly diminished. It is therefore becoming difficult to justify a legal concept of privacy purely based on the idea that what takes place in private environments involves privacy and what takes place in public does not.[71]

Nevertheless, the role of the private realm, and the role of information privacy law in protecting that realm, is still redolent in the context of preservations of informational access for personal autonomous growth. In this respect, there is a clear link between information privacy and the flourishing of autonomous individuals in liberal orders.[72] As noted earlier, there are some parallels between control concepts and their role in protecting individual autonomy. However, the focus of informational access for autonomous growth moves the conceptual discussion from decisions and mechanisms about control of information to the ability to limit access to private information. It is this ability which enables the preservation of private information that ensures reflective and personal space for personality development to take place.[73]

---

[68] Daniel J Solove, '"I've Got Nothing to Hide" and Other Misunderstandings of Privacy' (2008) 44 San Diego Law Review 745, 755.
[69] Daniel J Solove, 'Privacy and Power: Computer Databases and Metaphors for Information Privacy' (2001) 53 Stanford Law Review 1393.
[70] Helen Nissenbaum, Privacy in Context: Technology, Policy, and the Integrity of Social Life (Stanford Law Books 2010) 113.
[71] Helen Nissenbaum, 'Protecting Privacy in an Information Age: The Problem of Privacy in Public' (1998) 17 Law and Philosophy 559, 570.
[72] Steeves 325.
[73] Birnhack.

In a shallow sense, considerations of access to information form the flipside of control decision making, and information privacy thus becomes about decisions in relation to who has access to information.[74] However, in a much more profound and deep sense, informational access intrinsically relates to private spaces for autonomous growth in which the ability to build barriers[75] or send signals[76] as to a requirement for privacy is a cherished feature of the broader liberal ambit. The conceptual basis of informational spaces broadens conceptual considerations to autonomous individual choice.[77] The ability to allow and deny access to private information creates unfettered processes of individual decision making that symmetrically protect personal autonomy. Access to information, in this sense, also carries with it broader connotations of space, as there is an inherent recognition that limiting access to information also crosses over with limiting access to spaces where such information is generated. Accordingly, there is a more clearly articulated conceptual sense that information generation is a social practice that is embedded in the societal conditions of everyday life.

'Limited access to the self' concepts regard the requirement that humans must separate and conceal from each other certain parts of their lives, including their information, in order to flourish. This can be in the form of physical access (e.g. access to our own bodies), spatial access (e.g. access to our homes or private spaces) and informational access (e.g. access to our personal information). Access limitation, in this sense, is more than mere solitude, which is the ability to withdraw from other individuals and society.[78] Instead, limited access concepts are concerned with wider social complexities that are fundamental to liberal societies and set the boundaries for individual freedoms against collective interference and intrusion. A degree of inaccessibility is consequently an important protection against unwarranted intrusions into private life and reflects an important condition of human life – namely, that personal lives are protected by being able to place limits on who can access information, bodies and spaces.[79]

---

[74] Moore.
[75] Ibid 23.
[76] Stanley I Benn, *A Theory of Freedom* (Cambridge University Press 1988) 267.
[77] Austin, 'Re-reading Westin' regarding space development in infrastructure to 'secure meaningful choice'.
[78] Solove, 'Conceptualizing Privacy' 1103.
[79] Hyman Gross, 'The Concept of Privacy' (1967) 42 New York University Law Review 34.

Ruth Gavison provides one of the clearest discussions on why humans need to be able to limit access to themselves and their private lives. For Gavison, privacy regards

our concern over our accessibility to others: the extent to which we are known to others, the extent to which others have physical access to us, and the extent to which we are the subject of others' attention.[80]

Privacy therefore has certain fundamental concerns that relate to individual accessibility to other individuals and societal groups. It is what other individuals and groups can know about us, the extent to which they are able to physically intrude into our private lives to find out more details and thus the extent to which personal information can be used to draw individuals into a centre of attention that further illuminates our private lives. Privacy can therefore be gained in three independent but interrelated ways:

1. *Secrecy* (e.g. where no-one has information about you);
2. *Anonymity* (e.g. where no-one pays attention to you); and
3. *Solitude* (e.g. where no-one has physical access to you).[81]

Perfect privacy is thus attained when we are completely inaccessible to others. However, this is an unrealistic consideration, as the situations in which we can make ourselves completely inaccessible are limited in modern societies. The concept of a loss of privacy becomes more important, and a loss of privacy occurs 'as others obtain information about an individual, pay attention to the individual, or gain access to the individual'.[82] As such, the informational privacy component of access concepts also serves as a wider social component. Privacy, and information privacy, is essential to democratic government because it fosters and encourages the moral autonomy of the citizen. In other words, it provides the social space for individual development.

Total lack of privacy is full and immediate access, full and immediate knowledge, and constant observation of an individual. In such a state, there would be no private thoughts, no private places, no private parts. Everything an individual did, and thought would immediately become known to others.[83]

---

[80] Gavison 423.
[81] Ibid 428.
[82] Ibid.
[83] Ibid 443.

Gavison thus links informational access with physical and spatial access and thus begins a process of situating information privacy in a deeper social context than that of control-based concepts. This broader process specifically attempts to recognise the social and political ambits of information privacy and its role in personality development. Expansive privacy protections, even including those that form narrower information privacy concepts, play an integral role regarding individual integrity, dignity, the preservation of freedom and autonomy.[84] As Edward Bloustein explains,[85]

[a] man whose home may be entered at the will of another, whose conversation may be overheard at the will of another, whose marital and familial intimacies may be overseen at the will of another, is less of a man, has less human dignity, on that account. He who may intrude upon another at will is the master of the other and, in fact, intrusion is a primary weapon of the tyrant.[86]

Privacy as personhood accords a much wider notion of what privacy is, what it attempts to achieve and how it attempts to achieve it.

The injury is to our individuality, to our dignity as individuals, and the legal remedy represents a social vindication of the human spirit thus threatened rather than a recompense for the loss suffered.[87]

Intrusions of privacy consequently affect core human functions of personality and dignity. Other authors have argued that this personhood thesis really represents a boundary between the state and the individual.[88] In situations where identity construction or self-definition is at stake, the state should not interfere.[89] Privacy consequently involves the protection of identity formation. Privacy protections provide us a space to determine who we are, including spaces and decisions of informational access. Stanley Benn also examined the role that informational privacy plays regarding the protection of personal autonomy in liberal societies. In this situation, privacy is about respect for autonomy, based on terms of the individual's unfettered capacity to choose.[90] For Benn, informational control and

---

[84] Ibid.
[85] Edward J Bloustein, 'Privacy as an Aspect of Human Dignity: An Answer to Dean Prosser' (1964) 39 New York University Law Review 962, 973.
[86] Ibid 974.
[87] Ibid 1003.
[88] DeCew 20.
[89] Jed Rubenfeld, 'The Right of Privacy and the Right to Be Treated as an Object' (2001) 89 Georgetown Law Journal 2099.
[90] Solove, 'Conceptualizing Privacy' 1117.

access decisions are directly linked to individual and societal requirements for autonomously functioning individuals as a precept of liberal
societies.

Benn outlined his construct of autonomy and the role of privacy in
its protection in *A Theory of Freedom*. Benn argued that the autonomous person, the truly 'self-made man', is an intrinsic component
of the liberal tradition[91] – the embodiment of the rational individual
who can freely make self-serving choices in line with the broader values of their society. The ability to make unfettered choices is thus
a vital constituent in notions of freedom predominantly predicated
úpon non-interference with individual actions.[92] Non-interference of
choice-making decisions is at the core of Benn's tripartite construction, which distinguishes three different concepts – namely, autarchy,
autonomy and heterarchy. These different concepts operate on a hierarchical scale of non-interference and associated rational decision-
making capabilities.

Autarchy refers to the nominally or normally functioning human
being. In other words, the autarchic agent is the standard instance of a
rational human being.[93] Autarchy is thus judged in terms of rationality to
form a baseline comparator by being at the centre point of the hierarchy.
The autarch is thus the reasonable person to be found travelling on the
Clapham omnibus.[94] Autonomy, on the other hand, is a state beyond
autarchy. It is an ideal to which all humans strive, to varying degrees, and
thus it is not part of the nominal or normal condition.[95] The autonomous agent, unlike the autarch, lives according to a law that they prescribe for themselves. Living under such a self-defined scope leads to
certain actions that are broadly valued in the wider society, such as independence of mind over a rabid, unquestioning acceptance of authority.
Non-interference is a general precept for all humans, but on the hierarchical sliding scale, interference is more justifiable as a paternalist measure for individuals who are not capable of meeting the norm of autarchy,
and non-interference is the norm for the autonomous individual.

Heterarchy relates to domination.[96] It is the situation where interference is no longer necessary because the individual, the heterarchic

---

[91] Benn 10.
[92] Ibid 11.
[93] Ibid 155.
[94] Ibid 156.
[95] Ibid 155.
[96] Andrew Roberts, 'Why Privacy and Domination' (2018) 4 European Data
      Protection Law Review 5 for a recent account and focus of privacy as the prevention
      of domination.

agent, is programmed to follow another set of authoritarian values.[97] Benn's heterarch is thus an automaton of another rather than an autonomous or aurtarchic individual. Heterarchy is therefore a condition

in which a person's preferences, his beliefs, or his capacity to act on his belief commitments have been rigged or impaired by methods that intentionally circumvent or block his rational decision-making capacity.[98]

That condition destroys an individual's ability to conduct any meaningful questioning of the world around them without promulgating immense anxiety concerning what they see in their own culture.

[I]nstead of perceiving himself as a natural person originating action, he now sees himself only as behaving in ways apt to bring about determinable outcomes. What he formerly understood as the importance of goals meshed in a web of beliefs that constituted his own identity would now appear only as a kind of mechanical force attracting him to the realization of certain preset states of affairs. Action would no longer be the manifestation of his own creative agency.[99]

The autonomous individual, on the other hand, will undertake the opposite. They will critically reflect and situate their own beliefs within the wider belief system of their society and critically assimilate both to better understand their world.[100] The critical appraisal of the seemingly complex entanglements of life fuels the ability to critique that is required for autonomy to flourish.[101]

Benn's notion of freedom, and the role of autonomy promotion, as a precept for freedom inherently regards non-interference in relation to informational decision making. In the positive sense, it requires the autonomous or autarchic individual to act on their self-defined motivations or beliefs.[102] It also heralds the principled application of privacy as a protection of personal autonomy. To explain this Benn outlined different categories of privacy that cut across the private dimension and notions of control and access.

For example, Benn's state of privacy simply refers to a state of being private. In turn, states of privacy are dependent upon privacy powers which reflect an individual's ability or power to control access by

[97] Benn 167.
[98] Ibid.
[99] Ibid 167–8.
[100] Ibid 179.
[101] Ibid 180.
[102] Ibid 170.

others to a private object.[103] Similarly, a privacy interest emerges when an individual chooses a state of privacy or has a power to make something private. Private, in Benn's sense, is therefore a norm through which rights can be imposed on objects such as correspondence, affairs or rooms. The establishment of the private signals an expectation of appropriate behaviour in relation to objects that are so characteristically deemed.[104]

Benn acknowledged that his notions of autonomy and privacy are closely bound to the liberal ideal and the requirement for private objects/sanctums that foster autonomous growth.[105] Privacy intrusions are therefore unjustifiable interferences into the interests that autonomous individuals have in establishing, sustaining and developing personality and personal relations.[106] Like Gavison, Benn intertwines informational, spatial and attention interests so that the protection of personal information becomes the ability to prevent 'unauthorised access to facts about oneself ... that would impair one's capacity to manage the complex system of appearances'.[107] However, Benn's protection of personal information has a much broader and explicit autonomy-related perspective than the control concepts outlined earlier.

Benn also acknowledged that while the focus of privacy protections is autonomy, the construct of private, and thus privacy, is not within the sole discretion of the autonomous individual. Privacy interests are not solely defined by what autonomous individuals can perceive that they control. Instead, they will depend 'on rules of law, morals, or conventions, according to context'.[108] Benn was acutely aware of the contextual considerations and the impact these could have on classifications of private and thus privacy:

[T]he liberal cannot give absolute specifications...for what is private and what is not, because privacy is...relative to the social nexus in which it is embedded.[109]

As we will see in the next thematic extension, the idea of societal context and the intersubjective nature of privacy conceptualisation becomes important in contemporary concepts of information privacy.

---

[103] Ibid 266.
[104] Ibid 267.
[105] Ibid 268.
[106] Ibid 282.
[107] Ibid 288.
[108] Ibid 269.
[109] Ibid 271. Cited in Allen, *Unpopular Privacy: What Must We Hide?* 15.

5.4   The Social and Relational Context

We are starting to get a clearer idea of how different concepts of information privacy provide different emphases about what information privacy law should protect. Control concepts emphasise the requirement for individually oriented mechanisms, processes and rights. Informational access considerations promote private zones that allow the flourishing of autonomous individuals as a founding precept of liberal society. Control and access therefore represent the initial root structure of information privacy's conceptual entanglement with broader privacy concepts. However, even while these roots begin to take substantial hold, a new conceptual root system starts to develop that critiques both the individualised aspects of control and the liberal perfumery of the autonomy justifications of information privacy. This contemporary root system gathers from different directions but formulates broadly around a key idea – namely, that the practices of privacy and information privacy are situated in social and political contexts.[110] Contemporary concepts thus highlight the social and relational context of information privacy, which leads to the construction of new forms of critique,[111] including critical appraisal of the very process of conceptualisation of information privacy.

Solove provided a comprehensive critique of traditional conceptualisations of privacy that attempted to determine common elements unique to privacy within commonly understood notions of what is privacy, including information privacy as control over information, as noted previously.[112] Solove argued that a reconstruction of what is the very understanding of privacy is required and needs to be undertaken from a pluralistic, bottom-up perspective that views legal remedies as 'a set of protections against a cluster of related problems'.[113] A pragmatic approach to conceptualising privacy was necessary, because it would focus on privacy in specific contextual situations rather than presenting abstract conceptions.[114] Social context thus becomes a vital aspect of privacy conceptualisation, including concepts of information privacy.

Solove's argument is predicated upon this notion. He contended that privacy problems which arise from the increased use of information technologies will not be resolved by 'clinging to a particular conception

---

[110] Priscilla M Regan, 'Privacy and the Common Good: Revisited' in Beate Roessler and Dorota Mokrosinska (eds), *Social Dimensions of Privacy* (Cambridge University Press 2015) 53–5 provides a beautifully succinct overview.

[111] Bannerman 6 on the importance of relational theory that goes beyond liberal autonomy.

[112] Daniel J Solove, *Understanding Privacy* (Harvard University Press 2008) 14.

[113] Ibid 40.

[114] Solove, 'Conceptualizing Privacy' 1128.

of privacy', especially regarding the collection and use of personal information.[115] Accordingly, the conception of privacy as control over information only partially captures the problems that arise from increased use of personal information. The resolution of these problems required an analysis of the social problem, 'beginning with the problem itself rather than trying to fit the problem into a general category'.[116]

Solove highlighted that the social context of information generation and provision is a latent but ever-present component of the information privacy law literature. Other authors have also addressed this point even with the scope of control concepts and liberal orientations. Schoeman outlined that the wider concept of privacy is part of a 'historically conditioned, intricate normative matrix with interdependent practices' and is best understood when viewed contextually.[117] Allen contended that information privacy and social context are intimately bound with the creation, development and maintenance of social relationships.[118] Privacy is 'down time' that provides the space for reflection and thus allows individuals to prepare themselves for their wider social responsibilities within the context of their own lives.[119] Privacy as a social practice thus shapes individual behaviour in conjunction with other social practices and is therefore 'central to social life'.[120] Moor and Tavani, who favour a control approach, also acknowledge the importance of 'situations' in deciding when an individual has a condition that is equivalent to privacy.[121] However, the notion of a situation is characterised as 'deliberately indeterminate or unspecified' so that it can be construed in a number of different ways in circumstances that would normally be regarded as private.[122] In other words, social context is key.

The clearest conceptual statement of the importance of social context in the application of information privacy law is Helen Nissenbaum's

---

[115] Ibid 1151–2.

[116] Ibid 1154.

[117] Ferdinand David Schoeman, *Privacy and Social Freedom* (Information Privacy, Cambridge University Press 1992) 137.

[118] Allen 739–40 regarding the value of privacy that therefore lies in 'the context in which individuals work to make themselves better equipped for their familial, professional, and political roles'.

[119] Schoeman regarding the role of privacy in the balancing of social freedoms and an individual's need to be part of a 'human context'.

[120] Ibid 137.

[121] Moor 30 stating that privacy is normatively prevalent if an individual or group is protected from intrusion, interference and access by others.

[122] Tavani 10 explaining the role of Moor and Tavani's Restricted Access/Limited Control (RALC) theory.

Privacy in Context[123] and her concept of *Contextual Integrity*.[124] Nissenbaum put forward an analytical framework to examine potential privacy concerns arising from the introduction of new technologies or technological structures principally involving the use of personal information.[125] Privacy is sufficiently important to the continued existence of social and political life that it cannot be compartmentalised and reduced in social importance.[126] Instead, contextual integrity represents privacy as a 'delicate web of constraints' relating to flows of personal information that balances the multiple political and social spheres of human life. An attack on individual privacy is therefore an attack at the 'very fabric of social and political life'.[127] Privacy in this regard is not a claim regarding an individual's control of their personal information but rather entails a right to the appropriate flow of personal information which is systematically grounded in the characteristics of social situations.[128]

Contextual Integrity is founded on social context and gains expression through its primary concept, context-relative informational norms. These norms govern entrenched expectations that in turn govern flows of personal information in everyday life.[129] Accordingly, a breach of privacy under the theory of Contextual Integrity equates to a violation of an established informational norm.[130] These norms are characterised by four key parameters. *Contexts* provide a backdrop for norm development and feature an array of components[131] that abstractly represent the experienced social structures of everyday life.[132] *Actors* are those participants involved in the direct context of information exchange: senders and receivers of information and information subjects.[133] However, the type of relationship that each party has with each other is not fixed, and it is acknowledged that both individuals and organisational representatives can have different capacities in different situational

---

[123] Nissenbaum, *Privacy in Context*.
[124] Helen Nissenbaum, 'Privacy as Contextual Integrity' (2004) 79 Washington Law Review 119.
[125] Nissenbaum, *Privacy in Context* 6–7.
[126] Ibid 128.
[127] Ibid.
[128] Ibid.
[129] Austin, 'Re-reading Westin' 62 regarding norms of public behaviour of looking-away etiquette as a state of Westin's reserve.
[130] Nissenbaum, *Privacy in Context* 140.
[131] These are canonical 'activities, roles, power structures, norms (or rules) and internal values (goals, ends, purposes)'.
[132] Nissenbaum, *Privacy in Context* 134.
[133] Ibid 141.

circumstances.[134] *Attributes* refers to the type or nature of the information in question. For example, the same type of information can have different meanings or applications in different contexts. Finally, *transmission principles* provide a constraint on the flow of information from party to party in a given context by stipulating terms and conditions which govern the transfer of personal information.

These parameters are embedded within informational norms, which in turn are embedded within different social contexts.[135] As such, different parameters come to the fore in different social contexts and in the guise of different privacy-related problems. For example, in a context of information exchange amongst friends, there is expected transmission principles – namely that the personal information exchange is usually volunteered freely and there are certain trust-based expectations about how that information will or will not be used. However, the medium of exchange can impact upon friend-based transmission principles, especially in situations involving a broader and thus less controlled transmission of personal information. Likewise, the provision of the exact same personal information is likely to vary between the context of a patient-doctor relationship during a medical consultation and that of an interviewee-interviewer relationship in relation to an employment application. The analysis of informational norms and component parameters are best conceived as juggling balls that move in sync with different emphases placed on different balls depending on the social context involved and the privacy concern emanating therefrom.

Nissenbaum developed the theory of Contextual Integrity as a 'framework for determining, detecting, or recognizing when a violation [of information privacy] has occurred'.[136] To do so requires a comparison between entrenched and novel practices to adduce whether there has been a violation of context-relative informational norms. *Privacy in Context* is cemented the importance of contextualisation in the examination of concerns relating to the provision, protection and use of personal information. However, Nissenbaum has acknowledged that much work is yet to be undertaken about how Contextual Integrity can apply to existing information privacy legal regimes, especially comprehensive frameworks.[137]

---

[134] Ibid 143. Nissenbaum contends that an actor in one situation may not act in the same way as in another. For example, the difference between an actor in a 'businessman to employee' relationship compared to a 'parent to child' relationship. Accordingly, the capacity within which an actor may act has an 'innumerable number of possibilities'.

[135] Ibid 145.

[136] Ibid 148.

[137] Ibid 238. Nissenbaum suggests that her theory of contextual integrity may be more suited to sectoral frameworks because 'it embodies informational norms relevant to specific sectors, or contexts, in the law'.

Julie Cohen situates the relational context of both privacy and information privacy within social structures that give rise to selfhood development. Information privacy, in this regard, entails 'an interest in breathing room to engage in socially situated processes of boundary management'.[138] At face level, Cohen's notion of 'breathing room' could be seen to parallel both control and informational access concepts, as they claim to provide the very same spaces for personality development and autonomous flourishing. However, Cohen then makes clear that the focus and application of such breathing spaces departs significantly from the forms, but not necessarily the ideals of liberalism.[139] Part of the reason for this shift lies in the individualised form of legal protections accorded through information privacy rights. The conventional locus of foundational rights discourse, based on notions of liberalism, does not adequately reflect the subjective development of individual and societal privacy-related expectations that are formulated through relational, contextual and spatial social practices.[140]

Cohen's criticism of the liberal approach to privacy, and information privacy, as a protection of personal autonomy begets a radically different approach to selfhood development.[141] To understand what information privacy seeks to preserve, as an ameliorator of selfhood harms, it is first necessary to have a theory of selfhood and thus of the type of self-determination that privacy enables.[142] A problem arises because the legal academy, at least in the United States,[143] is so imbued with privacy as a liberal protection of autonomy that it has not articulated a sophisticated theory of the self beyond the confines of the readily assumed autonomous self. In other words, autonomy in the liberal sense resides as a separate state of being. The role of privacy and information privacy protections is therefore to preserve enough individual space for the state of autonomy to flourish. Freedom from interference is thus enough to generate states of individual and independent autonomous growth. However, Cohen argues that these highly individualised and independent autonomous states of being and selfhood do not accurately reflect

---

[138] Cohen, 'What Privacy Is For'. See also Steeves 341: 'privacy is the boundary that creates and protects the democratic relationships that are central to us as a community'.
[139] Cohen, *Configuring the Networked Self*, 29.
[140] Cohen, 'Turning Privacy Inside Out' 20.
[141] See also Austin, 'Re-reading Westin' 67.
[142] Cohen, *Configuring the Networked Self* 10. Note also the positive dimension in which privacy is expressed as 'freedom to' through enablement rather than a negative 'freedom from' interference.
[143] Ibid 109.

the complexities of networked information societies. Therefore, systemic attention is required to promote understanding of selfhood development as part of the 'bodies and spaces within which individuals and groups reside and ... the materiality of artifacts and architectures'.[144]

The underlying liberal notion of autonomy eventuates in a 'paradox' in the privacy context and does little to help explain the role that privacy and autonomy play in relation to each other. These issues are further complicated when considered through the lens of conflicting notions of autonomy from a negative and positive sense that still belie 'an autonomous core – an essential self-identifiable after the residue of influence has been subtracted'.[145] Consequently, thinking about privacy, and what it seeks to protect, requires thinking about selfhood development in a fundamentally different way, as selfhood 'comes into being and is negotiated through contexts over time'.[146] Privacy for Cohen is therefore

The privacy embedded in social practices of boundary management by situated subjects preserves room for the development of a critical, playful subjectivity that is always-already intersubjective – informed by the values of families, confidants, communities, and cultures. In a world with effective boundary management, however, there is play in the joints, and that is better than the alternative. And on this understanding, privacy implicates not only individual interests, but also collective interests in human flourishing and in the ongoing development of a vibrant culture. Privacy's goal, simply put, is to ensure that the development of subjectivity and the development of communal values do not proceed in lockstep.[147]

Cohen's critique of the liberal, autonomous self and how it is presumed to be created questions the role of privacy and information privacy and what they seek to protect. One could read the distinctions between the liberal and postmodern self as being irreconcilably and fundamentally different. However, Cohen explicitly highlights that the distinction in itself is a choice, and the more fruitful choice is to explore both selves as 'endpoints on a continuum along which social shaping and individual liberty combine in varying proportions'.[148] The role of privacy conceptualisation within this continuum 'should explain what function privacy performs in a world where social shaping is everywhere and liberty is always a matter of degree'.[149]

[144] Ibid 25.
[145] Ibid 113.
[146] Ibid 114.
[147] Ibid 150.
[148] Ibid 115.
[149] Ibid.

We now move on to the last conceptual theme and reconsider notions such as control and social shaping – not from the perspectives of individual abilities and social and relational contexts, but from that of power relations. Information privacy, in the final theme, therefore seeks to protect against the accumulation of power by data-collecting organisations that use different forms of power in different ways to achieve their ultimate ends.

5.5   The Structural Problem of Power

The coverage of the three previous themes all allude to differing degrees, to an underlying subtext of information privacy as a protector or an ameliorator, of power relations. The overt focus of individual control processes in the first theme begs the question of what it is that enhanced forms of individual control seek to achieve. As explored in previous sections, and covered again in Chapter 6, there is a belief predicated on the rational ability of individuals to make autonomy-reflecting decisions about information exchange. Control, in this regard, overlaps with the second theme and the enhancement of autonomy, which is a protector against totalitarian actions, by protecting both private spaces and thoughts.[150] But the need for control also suggests that there is a wider societal consideration, as outlined by some of the concept's key critiques and the relational context of information privacy as a social web or fabric. The broader societal dimension involves forms of power relations, such as Benn's heterarchy and Cohen's contentions about social shaping.

A power-related analysis therefore questions the very basis of the conceptual and legal application of information privacy against the imbalance of power between individuals and data-collecting organisations. Bennett and Raab, picking up on many of the points highlighted earlier, make a number of powerful contentions that involve the historical development of information privacy law.[151] One important aspect that emanates from their work is that the content and provision of information privacy protections is inherently dependent on the context of social application and is thus applied subjectively by individuals to their own circumstances.[152] Bennett and Raab situate the ongoing development of information privacy law within a much wider sphere than that assumed by the traditional information privacy paradigm founded upon liberal notions.[153] The general application of information privacy law

---

[150] Moore 24.
[151] Bennett and Raab.
[152] Ibid 9.
[153] Ibid 17.

consequently involves deeper societal disputations that involve power relationships in the form of class, gender and race as well as other social categories.[154]

Flowing from this analysis, it should be no surprise that the concept of information privacy as an ameliorator of power is controversial. It highlights the degrees of powerlessness that individuals possess in the face of dominant private and public sector data collectors. It contends that individuals are deliberately excluded from corporate decisions to reuse their personal information. It questions the capacity of government regulation and current laws to adequately protect individuals. The concept thus challenges the validity of information privacy's dominant paradigm – that individuals can and should maintain control over their personal information. Several contemporary critiques of information privacy highlight the contextual nature of social relationships and the role of the individual within the networked societies of the collected world.

Power in the information privacy law literature is often referred to as 'power over' – namely, the ability to exert power over another in order to achieve a desired aim – but the literature also refers to other concepts, such as hegemonic power and Foucauldian concepts of power, as the use of power without aim. Accordingly, whilst this concept of information privacy has critical merit in the literature, it currently lacks a principled conceptual base[155] that restricts further development, particularly in the context of the rapid technological development occasioned by the collected world. Some authors directly relate privacy concerns about surveillance issues with recognised sociological and critical works on power, but for most authors, the link is largely implicit and lacks foundational explanation.

One of the most influential works on information privacy as concern about power is Solove's work on the use of databases and digital dossiers that exacerbate existing power relations.[156] Writing in the early 2000s, Solove argued that the advent of the 'Information Age' had ensured that previously mundane pieces of personal information had a newfound value due to data collection advances and the enhanced use that such information can be put to. The prime target was commercial marketing databases and the effect they had on magnifying imbalances of power relationships as a result of what was later termed data being the new oil, leading to the advent of vast corporate database assets.[157] In turn, the

---

154 Ibid.
155 Roberts 11.
156 Solove, 'Privacy and Power' 1393.
157 Ibid 1408.

focus on these new forms commercial value provided greater impetus for personal information collection. The increased value of marketing databases became some early descriptors of big data logics and sensorised collections as commercial entities now felt the need to know more about the minutiae of individual lives.[158]

Solove, now famously, contended that the concept of information privacy and the legal concerns that flowed were based on an outdated but dominant paradigm of information privacy as a Big Brother problem. Another metaphor was therefore required to adequately conceptualise the power-related issues arising from ever-increasing data collection, storage and analysis strategies. Solove suggested that Kafka's *The Trial* rather than Orwell's *1984* was the better metaphor through which to view the conceptual perspective of information privacy as a power relationship between individuals and bureaucratic organisations of both the public and private sectors. Both metaphors portray information privacy problems as an exercise of power, but in different ways.

The concept of information privacy under the Big Brother metaphor represents the use of coercive power by governments to oppress, control and dominate. It involves the application of privacy-invasive techniques to maintain the power of a totalitarian state over its subjects via overt surveillance. Control is maintained through rule of fear and through constant monitoring. The Big Brother metaphor therefore

[u]nderstands privacy in terms of power, and it views privacy as an essential dimension of the political structure of society.[159]

In the context of personal information collection, surveillance methods under the Big Brother metaphor are devised through 'dataveillance',[160] leading to the systematic collection and use of personal and nonpersonal data by bureaucracies for surveillance purposes.[161] Dataveillance is a form of surveillance that is able to curtail individual freedom because individuals who know that their data is being watched will alter their behaviour. Control is therefore gained via the systematic collection and use of personal information that provides a detailed insight into an individual's life.

Solove argued that whilst the Big Brother metaphor was helpful to address the clearer information privacy issues relating to surveillance,

---

[158] Ibid 1396.
[159] Ibid 1415.
[160] Roger Clarke and Graham Greenleaf, 'Dataveillance Regulation: A Research Framework' (2017) 25 Journal of Law, Information and Science 104.
[161] Solove, 'Privacy and Power' 1417.

invasions of privacy and the inviolability of personal thoughts, the metaphor does not provide an adequately accurate perspective of the concerns generated by the mass collection and use of personal data through bureaucratic databases.[162] First, as highlighted previously, the Big Brother metaphor focuses largely on surveillance as a use of power to shape and coerce individuals. However, most personal information collections are not aimed at gaining control over a populous, but are intended to know more about us as a means of studying and exploiting our expressions of individuality rather than suppressing them.[163] Second, much of the personal information that is collected by corporate and governmental bureaucracies does not have an embarrassing element. Finally, bureaucratic personal data collection is conducted by a myriad of 'Little Brothers' for a wide range of purposes rather than by one omnipotent government agency for one purpose.[164] Consequently, information privacy concerns do not just relate to the direct exercise of governmental power to collect and reuse personal data to monitor, control and oppress individuals.

Solove argued that the Kafka metaphor offers a more realistic analysis of the information privacy concerns relating to databases and the power issues entailed. The primary information privacy problem with databases stems from the way the bureaucratic process treats individuals and their information, such as the degree of ennui and lack of interest that Joseph K faced in *The Trial*. Bureaucratic and corporate databases do not create power-related problems, but they do exacerbate existing difficulties because they magnify power imbalances that transform existing relationships in new ways which impact upon personal and societal freedoms.[165] The real privacy problem that emerges relates to an individual's lack of control over the use of their personal information by faceless bureaucracies which thus eliminates individual participation in fundamental life and societal decisions. As such, information privacy problems can occur from a group of disempowering practices associated with the collection and use of personal information.

A precondition for successful information privacy regulation must be to establish rules that govern the power relationships between individuals and bureaucracies.[166] Such rules should seek to equalise power imbalances and thus ensure the instigation of fair, voluntary and informed information transactions. This goes beyond traditional notions of information privacy that focus on the control

---

[162] Ibid 1418.
[163] Ibid 1419.
[164] Ibid 1421.
[165] Ibid 1424.
[166] Ibid 1455.

of information and have been shaped within a property rights paradigm revolving around notions of ownership of personal data, as noted earlier.[167] The underlying foundations of information privacy law's control basis are no longer suitable for resolving current and future problems because of the dominant paradigms of surveillance and ownership which continually divert attention away from the real problem – the imbalance of power relationships.[168] The development of information privacy laws must therefore focus on the structure of power in modern society.[169]

[Privacy] is a problem that involves power and the effects of our relationship with public and private bureaucracy – our inability to participate meaningfully in the collection and use of our personal information. As a result, we must focus on the structure of power in modern society and how to govern such relationships with bureaucracies.[170]

The structural context of society and the power role that the private sector has over individuals because of that structure is therefore a key point of examination. The problem lies not with the notion of privacy itself but with societal and individual angst aimed at underlying imbalances of power between individuals and organisations. Power is the real thing that bothers individuals, but it is undefined and under-articulated in the privacy literature, as Rosa Ehrenreich argued.

The problem with the concept of 'privacy,' however defined, is that whatever work it does, its problematic overuse can obscure certain issues of power and consequential harm.[171]

The structure of modern society and the role of private sector service providers within that structure are integral to understanding the real power concerns relating to privacy issues. Many problems presented as privacy issues are in reality concerns of power dressed up as privacy problems, and there has been amongst lawmakers a 'readiness to lose sight of the issue of power and subsume it wholly into the murky and contested notion of privacy'.[172] The reason for this subsuming is that it is easier for policy makers to define, develop and contest arguments

---

[167] Ibid 1446.
[168] Ibid 1431.
[169] Ibid 1461.
[170] Ibid.
[171] Rosa Ehrenreich, 'Privacy and Power' (2001) 89 Georgetown Law Journal 2047, 2052.
[172] Ibid 2055.

about better privacy protections than to address the imbalance of power relations between individuals and corporations.[173] It would be more accurate to view privacy and power issues as a continuum of overlapping circles that represent different forms of privacy concerns.

Ehrenreich further examined why issues of power are in fact construed as issues of privacy, especially as both concepts are intimately bound together. She contended that power has not been discussed in tandem with privacy because of the imprecise nature of power, particularly in the form of Marxist discourse that has largely been discredited in the United States.[174] In effect, power is hard to talk about, but privacy is not, because 'the notion of privacy resonates well in a country so heavily seduced by the notion of "individual freedom"'.[175] Moreover, it is difficult for the American political discourse to distinguish fully between privacy and power because both concepts are so intimately bound together.

[I]t would probably not be an exaggeration to say that without privacy, power could not sustain itself; and without power, privacy could not exist.... [T]he realm of the 'private' is always constructed in relation to social power: Power constructs privacy and, to maintain itself, power also destroys privacy. Privacy, in turn, both constructs power and challenges it.[176]

The idea that power is involved in the construction of societal and individual identity leads into privacy as an issue of hegemony and the ability of dominant organisations to shape perceptions. Marcy Peek contended that there is a fundamental dynamic of information privacy laws that had remained largely unrecognised – namely, the power of private sector corporations to shape the development of laws to their own favour.

Information privacy laws have the effect of facilitating greater collection, sharing, and use of personal information because information privacy is governed as much by corporate actions and corporate decision-making as by governmental regulation. In other words, information privacy is governed not just by governmental law but also by corporations.[177]

The private sector defines information privacy laws, and legal development is not purely governed by 'governmental law'.[178] This reflects the

---

[173] Ibid 2058.
[174] Ibid 2057.
[175] Ibid.
[176] Ibid 2058.
[177] Marcy E Peek, 'Information Privacy and Corporate Power: Towards a Re-Imagination of Information Privacy Law' (2006) 37 Seton Hall Law Review 127, 137.
[178] Ibid.

fact that practical governance in society is now conducted through an interlocking web of public and private sector bodies that effectively define the boundaries of social existence and manage the expectations of individuals.[179] In this governed environment, personal information is viewed by corporations as a commodity that has enormous economic value and provides fuel for market-based economies. It is not surprising, then, that corporations drive the continual erosion of personal information privacy and have such an interest in the development of favourable information privacy laws.[180]

Legal development of information privacy gives the appearance of being in the hands of government, but corporate actors create and shape legal regimes which resemble government law making and enforcement. Corporations affect the law in ways that produce effects which appear to be governmental action but are in fact examples of corporate private governance.[181] The dynamic interconnectedness of the private sector in governmental law making reflects a symbiotic dance where one party acts and reacts to the other behaviours, decisions and regulations. As such, information privacy laws deflect individual and societal attention from the hegemonic forces that control those laws.

Indeed, information privacy law and enforcement are merely an exercise in legitimization, i.e., a cover or masquerade for corporate domination of the American consumer.[182]

This has profound implications for information privacy laws, because it becomes clear that truly effective solutions will only be generated by addressing the underlying power dynamic between individuals and corporations.[183] In terms of information privacy law, corporations are power centres and form the location of power struggles over the use of personal information. It is necessary to strip away the political façade of legitimisation and to reveal the hidden sources of power before issues regarding the rights of individuals can realistically be brought to the forefront.[184] Information privacy problems will not be solved unless these power foundations are exposed to show the symbiotic governance of both corporations and governmental entities.[185]

---

[179] Ibid 146.
[180] Ibid 138.
[181] Ibid 141.
[182] Ibid 138.
[183] Ibid 165.
[184] Seda Gurses and Joris van Hoboken, 'Privacy after the Agile Turn' <https://osf.io/preprints/socarxiv/9gy73/> 21 discussing Agre's construct of 'capture' and the reorganisation and optimisation of user activities.
[185] Peek 160.

However, not all power related analysis are so foreboding. David Lindsay has argued that power is not purely about negative applications and associations of repression or domination. Building on Foucault's contention that power can also have a positive effect, Lindsay argued that the concept of information privacy should be

[i]nterpreted in the context of progressive micro-political struggles over individual identity which occur within overall social processes of rationalisation and normalisation.[186]

Under a Foucauldian perspective, an analysis of the conceptual foundations of Australian privacy laws revealed an environment of complex struggles regarding the subjective creation and development of individual identity.[187] Information privacy was the focal point for political struggles regarding identity, and such struggles take place via the legal rights and obligations accorded by privacy laws. Such laws are a litmus test regarding legal protections of individual privacy and thus the orientation of contemporary legal systems. However, power struggles over individual identity do not represent liberation from a malevolent oppressor, but instead merely highlight the ongoing struggles over identity. In this sense, there is no privacy battle to be won, as such, but rather a recognition of the key societal role that privacy plays in protecting an individual's ability to self-define despite the paradoxical entrenchment of universal notions of self that restrict individual conceptions of self-identification.[188]

The concept of information privacy is therefore equally complex as the concept of power, because they both involve the impersonal processes of social ordering alongside the ability of individuals to create their own identity.[189] Privacy should be viewed in context with techniques of power that are dispersed in society and have diverse forms that are difficult to define. In this sense, it is not surprising that different concepts of privacy have been developed to explain different power concerns in different social contexts.

If power relations are everywhere, then privacy, which must be seen in the context of such relations, is an understandably diffuse concept, capable of multiple meanings.[190]

---

[186] David Lindsay, 'An Exploration of the Conceptual Basis of Privacy and the Implications for the Future of Australian Privacy Law' (2005) 29 Melbourne University Law Review 131, 178..
[187] Ibid 142.
[188] Ibid 143.
[189] Ibid 139.
[190] Ibid.

Under this approach, data processing is conceived as a threat to societal and legal rights accorded to an individual as a self-determining, autonomous and moral agent.[191] Mass personal data processing can provide the means for totalitarian practices because it makes possible the marginalisation of nonconforming individuals from the necessities and services of modern-day life. Moreover, whilst mass data processing has the capability of increasing bureaucratic efficiency, it does so via a deeply impersonal social practice in which individuals essentially become the instrumental means of bureaucratic action. The effect is to subsume the individual into the impersonal bureaucratic process of data processing, which eventually has a normalising effect.[192] In this sense, the idea of information privacy as power refers to data processing as a form of ongoing surveillance of individuals which fosters certain types of social conformity – namely, dataveillance as a form of institutionalising self-censorship.[193]

impersonal social processes of rationalisation and normalisation, which are undermining individual autonomy, and not at particular invasions of privacy by the state or other individuals.[194]

Many of the foregoing arguments are also redolent in the contemporary information privacy and privacy literature that has most recently manifested in arguments against surveillance capitalism.[195] The core arguments are only introduced at this juncture, because Part III's analysis of Julie Cohen's work will cover underlying issues in greater depth alongside the legal analysis of the collected challenges. The central arguments against surveillance capitalism relate to many of the issues already covered in this book, beginning with the construction of systems of ubiquitous smartness in Chapter 2, the technological frameworks of the smart home and their concomitant business models in Chapters 3 and 4. Key to all of these considerations, and indeed the broader power-related concepts highlighted here, is the manifest collection, storage and analysis of sensorised data and the ability to draw individual and population-wide inferences from such data. Again, the work of Julie Cohen is instrumental in understanding the broader power considerations that arise from the context of surveillance regarding the data analytic practices of the private sector.

---

[191] Ibid 161.
[192] Ibid 163.
[193] Ibid 162.
[194] Ibid 163.
[195] Shoshana Zuboff, 'Big Other: Surveillance Capitalism and the Prospects of an Information Civilization' (2015) 30 Journal of Information Technology 75; Karen Yeung, 'Five Fears about Mass Predictive Personalization in an Age of Surveillance Capitalism' (2018) 8 International Data Privacy Law 258.

Throughout her work, Cohen is at pains to differentiate and reconsider the complex relationship between information privacy and surveillance. Recently, she has argued that manifestations of power emerge from the corporate data analytic context as a modulating form.[196] This notion regards a theme redolent in the surveillance studies discipline relating to the ability of information flow control as a means to modulate individual behaviour for profit generation.[197] Cohen defines modulation as 'a highly granular, feedback-driven approach to the study, organization, and ongoing management of populations of consumers'.[198] Modulation is therefore considered a process of power involving information flow about individuals that primarily operates in two ways. First, it recasts individuals as data doubles in preparation for commercially targeted predictive and prescriptive analytical outcomes.[199] Second, and more importantly, it shapes the process of selfhood development by privileging 'economic conditions in which network users are alienated from the process of shaping'.[200] As in the hegemonic effect (highlighted previously), the modulators, powerful data collecting organisations, are then able to evoke 'our participation in our own construction as cultural subjects' in a process in which data generation of self-surveillance practices becomes the norm.[201]

Data generation about the self, once deeply personal and inherently private, now becomes a factor for profit maximisation on an enormous scale – a scale that can flick casually from the personal to the political, from knowledge to truth and from self-determination to owned determination. Behind these modulated outputs there also lies another form of power, one over knowledge that is inherent in the process of data analytics – namely, equating pattern recognition with claims of truth.[202] The construct of the pattern is thus itself one that is 'systematically infused with a particular ideology'. The much-vaunted objectivity of the data analytic output is therefore nothing but a 'deeply internalized, system of values'.[203] This system of values is important in the power context, because it is 'calculative, instrumental and unreflective'. In other words, modulation has the components that produce the heterarchic state that Benn warns cautiously of.

---

[196] Cohen's work on modulation is covered in Chapter 8.
[197] José van Dijck, 'Datafication, Dataism and Dataveillance: Big Data between Scientific Paradigm and Ideology' (2014) 12 Surveillance & Society 197.
[198] Cohen, 'Between Truth and Power' 8; see also 5 regarding the relationship between 'seamless, convenient personalisation' for consumers and 'efficient identification of high value consumers and more accurate projection and valuation of risk'.
[199] Cohen, 'The Biopolitical Public Domain' 14.
[200] Cohen, 'The Surveillance-Innovation Complex' 213.
[201] Cohen, 'Between Truth and Power' 65.
[202] Ibid 64.
[203] Ibid 65.

The genesis of Cohen's considerations can be tracked through the often-understated considerations of power in the information privacy literature. The warnings of modulation are reflective of Solove's digital dossiers and the different power contexts that could arise. Power in the modulated sense bears more witness to the subtler and often-underestimated uncaring of authority of the Kafka metaphor, as opposed to the deliberate oppressive goals of Big Brother. Similarly, Cohen highlights the continuing reluctance of the legal academies to fully address the issue of power, as Ehrenriech has also highlighted. Cohen's work finds traces in previous considerations of power, but it also highlights some key aspects of power-related critique that encapsulate surveillance capitalism,[204] as most clearly articulated by Shoshana Zuboff.

Surveillance capitalism aims to predict and modify human behaviour to produce revenue and market control.[205] The means of production thus serve the means of behavioural modification. The serving of behavioural modification also gives rise to a distinct form of power-entitled 'instrumentarianism' which is the

instrumentation and instrumentalization of behavior for the purposes of modification, prediction, monetization, and control.[206]

Instrumentarianism has two power-related components. Instrumentation refers to 'the ubiquitous connected material architecture of sensate computation that renders, interprets, and actuates human experience'.[207] Zuboff here is referring to the type of sensorised framework detailed in the anatomy of the smart home, discussed in Chapter 3. Instrumentalisation regards the passive and active social relations that translate and transfer the machine focus of surveillance capitalism into human society. In other words, instrumentalisation relates to the actions of human agency that provide knowing authority and unknowing passivity for the sensorised and machine learning frameworks that fuel behaviour modification.

Like Solove, Zuboff also distinguishes instrumentarianism as a new form of power separate from that of totalitarianism, even though the platform goliaths provide a gossamer-like appearance of the omnipotent

---

[204] Karen Yeung, 'Algorithmic Regulation: A Critical Interrogation' (2018) 12 Regulation & Governance 505, 514.

[205] Shoshana Zuboff, *The Age of Surveillance Capitalism: The Fight for a Human Future at the New Frontier of Power* (1st edn, PublicAffairs 2019).

[206] Ibid 6341, p. 352.

[207] Ibid.

type of control envisaged by Orwell. Zuboff thus replaces concerns relating to Big Brother with the development of the 'Big Other':

[I]t is the sensate, computational, connected puppet that renders, monitors, computes, and modifies human behavior.

Big Other combines these functions of knowing and doing to achieve a pervasive and unprecedented means of behavioral modification.[208]

Suffice it to say for the moment that surveillance capitalism augments, rather than replaces, the considerations of power highlighted in this section. We will return to some of the finer considerations of modulated power in Chapter 9 to consider information privacy law responses for a collected world.

5.6   Conclusion

This chapter has highlighted the different, yet overlapping, concepts of information privacy. It has attempted to show that different concepts of information privacy seek to protect different aspects of information exchange or flow. Control concepts seek to preserve individual control aspirations as part of regulated personal information exchanges. The second conceptual theme critiqued the highly individualised basis of control concepts and sought to demonstrate that information privacy has a wider purpose that seeks to secure informational spaces and thus protect personal autonomous growth. The third theme shifts the focus of information privacy from individual protections to a wider societal and relational context that seeks to protect the integrity of informational flows and intersubjectively modelled selfhood. This third theme highlights the broader societal context of information privacy through the often-under-articulated ameliorator of power. Recent scholarship has highlighted new notions of information privacy, including power-related analysis such as Cohen's construct of modulation and Zuboff's notion of instrumentarianism.

Part III of the book will return to and consider some key points of the conceptual analysis, particularly the power-related issues, of the collected world. In doing so, it will examine the conceptual foci of information privacy alongside the legal challenges that arise from the smart home and the smart home insurance business models. It is therefore time to shift the consideration from what information privacy seeks to

---

[208] Ibid 376.

protect to how information privacy law protects. We will thus examine information privacy law's series of individually focused, and principled, protections that provide opportunities for individual points of control in the processes of personal information exchange. These protections are also accorded through the imposition of procedural and substantive legal obligations that place limits on what can be done by collectors and users of personal information. It should also be increasingly clear why the control concept of information privacy was earlier defined as dominant, as information privacy law is very much focused on individual protections that foster procedurally intensive control.

6   How Information Privacy Law Protects

6.1   Introduction

This chapter follows on from the conceptual overview provided in Chapter 5 to examine the transition from information privacy as a concept to its legal manifestation. Like the conceptual analysis, a legal examination of how information privacy is applied is not a straightforward task. The transitional roots of information privacy law are historically grounded in the first conceptual theme discussed in Chapter 5 – namely, control concepts of information privacy. Information privacy law thus seeks to provide individuals with varying degrees of control and involvement in personal information exchange processes. However, information privacy law, whilst predicated on underlying notions of individual control, is also cognisant of the requirements of data-collecting organisations and the flow-on benefits of personal information use for society.

Accordingly, where control concepts of information privacy tend to focus predominantly on individual protections, the application of information privacy law tends to focus on the balancing of individual protections against organisational exigencies. Balance between individual protections and organisational requirements is thus a key component of information privacy law. However, analytical complexities arise because different jurisdictions have varying notions of how to conduct the balancing act and in so doing have different emphases about the weight and type of individual protections that should be accorded. As we will see in Part III, and as outlined in the book's opening chapters, the interconnectedness of the collected world does not necessarily respect jurisdictional boundaries, and data collectors for any smart home can effectively be based in any jurisdiction. Collections of sensorised data, particularly smart home event data, can thus transcend nationally focused legal systems. Challenges therefore arise in data collection structures that cut across the jurisdictional boundaries of different information privacy law systems.

Hence, we begin to consider some of the key complexities of information privacy law application in a collected world. These complexities run deep and are baked into the historical development of information privacy law in Europe, the United States and some of the non-EU Organisation for Economic Co-operation and Development (OECD) countries. The control basis of information privacy law is evident across all jurisdictions and is visible in legal structures that regulate the collection, storage and use of certain categories of information. The guiding form of control mechanism is privacy principles, otherwise known as fair information principles, that govern data-handling obligations for data collectors and provide a range of interaction points of involvement for individuals.

However, whilst the genesis of information privacy law regards similar core constructs, the path of implementation has diverged along different lines. The European model of information privacy law, particularly in relation to EU legal frameworks, places greater rights-based protections for individuals. Information privacy protections are foundational rights of EU citizenship. The US model of information privacy law places greater emphasis on market-based activities and therefore provides a lesser degree of protection for individuals.[1] The OECD-based systems place greater emphasis on balancing interests to facilitate data exchange processes, and thus any notion of rights-based protections exist in statutory rather than fundamental forms.

These different overarching regulatory perspectives influence the core constructs of information privacy law. As such, whilst the constructs are similar in concept across the three systems, they are different in application. We will focus attention on two crucial aspects of regulation that are pertinent to all three frameworks: regulated information and specific collection protections.

As to the first aspect, the type of regulated information in the three jurisdictions is different, both in terms of conceptual basis and intended regulatory reach. The effect is either a deliberately broad application, such as in the construct of personal data in the EU General Data Protection Regulation (GDPR), or a constrained application, such as in the different forms of personally identifiable information in the United States and the construct of personal information in Australia. The effect of which data is regulated or not is a crucial question for the application of information privacy law, as it has a primary and irreducible impact

---

[1] Paul M Schwartz and Karl-Nikolaus Peifer, 'Transatlantic Data Privacy Law' (2017) 106 Georgetown Law Journal 115.

on the degree of legal coverage which then flows.[2] Any information that is not regulated by different forms of information privacy law does not fall within the ambit of the law's protective framework.

Similarly, the second aspect – the application of the principled approach to information privacy – also differs across the three jurisdictions, particularly in relation to the availability of specific collection protections.[3] Both the EU and the OECD nation-based laws place obligations on data collectors at the point of collection. These obligations, in the form of specified collection or purpose specification principles, ensure that collected personal information is purposefully relevant to the data-collecting entity and that data collections are conducted in a fair and lawful manner. The US regime underplays collection requirements and places much greater emphasis on a procedural notification mechanism popularly entitled 'notice and consent'.

The requirement for collection obligations is proving to be an increasingly controversial point, as some commentators now audibly suggest that collection obligations should be nullified as a consequence of the drive for data ubiquity.[4] In other words, collectors should be able to collect any form of data, including personal information, without concomitant accountability obligations placed on collection processes, particularly in relation to an abstract notion such as fairness. Information privacy law should then focus on the real point of contention in the process's life cycle – namely, subsequent uses or disclosures, where harm quantification is most visible and risk assessment therefore becomes a viable regulatory mode in a world of big data. This book eschews analysis of the arguments in favour of regulating personal information use as an override of personal information collection to focus on the softer target of consent acquisition through privacy policies. It does this to demonstrate the foundational weaknesses of information privacy law's control model in a collected world. That said, however, Chapter 9's proposed legal reforms clearly outline the need for collection principles, and so the book finds an obvious place in this debate.

---

[2] T Zarsky, 'Privacy and Manipulation in the Digital Age' (2019) 20 Theoretical Inquiries in Law 1, 167.

[3] As noted in Chapter 1, and to repeat here, the focus of this chapter's analysis is on the applicability of collection principles as part of the broader analysis of life-cycle protections. This issue itself is controversial and needs deep consideration in relation to the collected issues that arise from sensorised collections.

[4] Tene and Polonetsky 239. See also Fred H Cate and Viktor Mayer-Schönberger, 'Notice and Consent in a World of Big Data' (2013) 3 International Data Privacy Law 67; Craig Mundie, 'Privacy Pragmatism' (2014) 93 Foreign Affairs; Mark MacCarthy, 'In Defense of Big Data Analytics' in Evan Selinger, Jules Polonetsky and Omer Tene (eds), The Cambridge Handbook of Consumer Privacy (Cambridge University Press 2018).

The differences in how regulated information is constructed and the differing ways in which collection obligations are treated become crucial to understand in the collected world. A taster insight into how these different effects will unfold has already been provided in Chapter 4 with the initial analysis of different smart home insurance models and how they treat regulated information. Chapter 7 will consider these effects in greater detail and will then outline the collected challenges arising from the collection of smart home event data.

In the meantime, the final subsection of this chapter returns to a core commonality of information privacy law in all three legal systems, which, again, varies in effect according to the application of regulated information and the existence of a principled protection for collection processes. Daniel Solove has termed this commonality 'privacy self-management', and it is reflective of a core underlying logic of information privacy law.[5] That logic places significant emphasis on the ability of data-providing individuals to make rational and informed decisions about the handling of their personal information, especially so in decisions regarding consent for collection. The discussion on privacy self-management returns to Chapter 5's conceptual coverage, which involves critical questioning of the control concepts of information privacy and information privacy's ability to foster personal autonomy. However, before we get that far, it is first necessary to cover a bit of history about the development of information privacy law frameworks, to tease out the application of information privacy law regarding the information it regulates and the principles used to create an obligatory foundation for data collectors.

6.2   Different Regulatory Perspectives

Chapter 5 labelled the control concept as the dominant conceptual paradigm of information privacy, the principal reason being that the control concept has been used as the basis for information privacy legislation and the development of what we now recognise as data protection or information privacy laws.[6] Westin's originating work in *Privacy and Freedom* was thus highly influential in its eventual realisation in the form of different laws across the globe.[7] Three legal instruments, developed in the 1970s and 1980s, have been essential to the development

---

[5] Solove, 'Privacy Self-Management and the Consent Dilemma'.
[6] Bennett and Raab 8 commenting that the policy problem of 'privacy' settled on the concept of information privacy. Chapters 7 and 8 of this book critique the control model.
[7] González Fuster 27–33 for an excellent overview of developments and historical application in general.

of information privacy law, and different laws share similar foundations based on the notion of a principled approach to privacy regulation.[8] However, different jurisdictions – particularly the United States, the EU and the OECD countries – have adopted different regulatory approaches to the governance of information privacy law issues. The United States incorporates a sectoral approach to information privacy, whereas the EU and the OECD countries have adopted a comprehensive approach.[9] These different approaches underlie dissimilar political requirements about the role and need for information privacy law. The EU greatly values the role that data protection has within contemporary European society, whereas the United States has traditionally opted for information privacy laws with less coverage and limited protections.

6.2.1   The Three Founding Instruments

The first founding legal instrument is the 1973 US Department of Health, Education, and Welfare (HEW) report entitled 'Records, Computers, and the Rights of Citizens'.[10] The HEW report's central issue involved the relationship between individuals and record-keeping organisations, which was increasingly being disrupted by the onset of growing processes of computerisation in public and private sectors. It was one of the first governmental attempts to find a balance between the organisational benefits arising from the enhanced efficiencies of automated personal information processing and the potential infringement of personal liberties from impersonal forms of data collection.[11] The balance was achievable through the concept of mutuality and by providing a degree of individual control over the collection of, access to and disclosure of an individual's personal information.[12]

The HEW report concluded that existing US laws of the early 1970s provided inadequate protection of individual privacy against potential

---

[8] Rules 25–27.

[9] Joris van Hoboken, 'From Collection to Use in Privacy Regulation? A Forward-Looking Comparison of European and US Frameworks for Personal Data Processing' in Bart van der Sloot, D Broeders and E Schrijvers (eds), *Exploring the Boundaries of Big Data* (Amsterdam University Press 2016) 242.

[10] Report of the Secretary's Advisory Committee on Automated Personal Data Systems, *Records, Computers and the Rights of Citizens* (United States Government Printing Office 1973).

[11] Robert Gellman, 'Does Privacy Law Work?' in Philip Agre and Marc Rotenberg (eds), *Technology and Privacy: The New Landscape* (Information Privacy, MIT Press 1997).

[12] Report of the Secretary's Advisory Committee on Automated Personal Data Systems 284. See also Birnhack 265.

record-keeping abuses and recommended the establishment of a federal 'Code of Fair Information Practice' for all automated data systems.[13] The report's recommendations led to the enactment of the Privacy Act 1974 (US),[14] which established the recommended Code of Fair Information Practice for Federal Government Agencies. The HEW report recognised the difficulties inherent in providing specific rule-based systems for personal information management, given the significant number of government agencies involved in data collection. The report's Fair Information Practice Principles (FIPPs) recognised that it was more preferable to deter inappropriate institutional practice than to mandate specified practices which had to be followed.[15] The original set of FIPPs are important because they laid a foundation for a principled approach to information privacy that was also adopted in other jurisdictions, thus establishing the principled core of information privacy law protections. These five core principles of fair information practice recommended by the report were as follows.

1. *Notice/awareness principle:* required organisations to give an individual clear notice about information practices before personal information was collected
2. *Choice/consent principle:* provided an individual the opportunity to consent to secondary uses of their information
3. *Access/participation principle:* ensured that an individual was able to access data about themselves to confirm that personal data held was accurate and complete
4. *Integrity/security principle:* obliged a data collecting agency to take reasonable steps to ensure that the data was accurate and was held in a secure environment
5. *Enforcement/redress principle:* provided an individual with the means to enforce a breach of the principles

In Europe during the same period, the Council of Ministers of the Council of Europe adopted two resolutions that concerned the protection of individual privacy arising from personal information held in private and public sector databases.[16] The resolutions were the instigator of a

---

[13] Gellman 195.

[14] Privacy Act of 1974, 5 § U.S.C. 552a.

[15] Katherine Jo Strandburg, 'Monitoring, Datafication and Consent: Legal Approaches to Privacy in the Big Data Context' in Julia Lane and others (eds), *Privacy, Big Data and the Public Good: Frameworks for Engagement* (Cambridge University Press 2014).

[16] Council of Europe, Committee of Ministers, Resolution 73 (22) on the Protection of the Privacy of Individuals vis-à-vis Electronic Data Banks in the Private Sector; Council of Europe, Committee of Ministers, Resolution (74) 29 on the Protection of the Privacy of Individuals vis-à-vis Electronic Data Banks in the Public Sector.

more substantial legal instrument to ensure adequate individual protections whilst enhancing the free trade of member countries. In 1981, the Council of Europe formally adopted the Convention for the Protection of Individuals with Regard to Automatic Processing of Personal Data,[17] which extended the ambit of the previous Council resolutions. The convention was intended as a catalyst to encourage and guide state legislative initiatives rather than provide a readily implementable set of data protection rules and regulations,[18] as exemplified by the generality of the convention's principles. Those principles required that personal information be

- collected and processed in a fair and lawful manner
- only stored for specified purposes
- only used in ways that are compatible with those specified at the point of data collection
- adequate, relevant and not excessive in relation to the purpose of data collection
- accurate and where necessary kept up to date
- preserved in identifiable form for no longer than is necessary
- kept adequately secure
- accessible by individuals, who have rights of rectification and erasure

Convention 108 acted as a template for the development of information privacy laws throughout the globe,[19] including further developments in Europe. Some fourteen years after the implementation of the convention, the then European Community (EC) adopted the Directive on the Protection of Individuals with Regard to the Processing of Personal Data and on the Free Movement of Such Data,[20] to create an EU-wide regime that set personal information governance standards for member states to follow.[21] The Data Protection Directive, as it was called, was a highly influential development within the broader

---

[17] Council of Europe, Committee of Ministers, Convention for the Protection of Individuals with Regard to Automatic Processing of Personal Data (1981).

[18] Lee A Bygrave, *Data Protection Law: Approaching Its Rationale, Logic and Limits* (Information Privacy, Kluwer Law International 2002) 34.

[19] Graham Greenleaf, *Balancing Globalisation's Benefits and Commitments: Accession to Data Protection Convention 108 by Countries Outside Europe* (UNSW Law Research Paper No 16-52, 2016).

[20] Directive (95/46/EC) on the Protection of Individuals with Regard to the Processing of Personal Data and on the Free Movement of Such Data [1995] O.J. L281/31.

[21] Lee A Bygrave, *Data Privacy Law: An International Perspective* (Oxford University Press 2014).

fostering of a data protection jurisprudence in the EC. However, it also had a foundational weakness. As a directive, the ultimate implementation of the Directive was left to member states. Therefore, different member states interpreted the Directive's scope in differing ways, to the extent that data protection application across the EU was not perceived to be uniform.

The Directive, as the primary legislative vehicle of the EU, became increasingly problematic in the face of charter and treaty ratification that increasingly enshrined data protection as a foundational right of EU citizenship. For example, Article 8 of the EU Charter of Fundamental Rights came into force in 2009[22] and provided a right of protection of personal data in Article 8(1),[23] further reiterating this right in Article 16(1) of the Treaty for the European Union. Article 8(2) of the Charter also added a specific reference to information privacy as a foundational protection of EU citizenry:

Such data [personal data] must be processed fairly for specified purposes and on the basis of the consent of the person concerned or some other legitimate basis laid down by law. Everyone has the right of access to data which has been collected concerning him or her, and the right to have it rectified.

The clamour to replace the Directive grew, given the focus of data protection as an EU fundamental right.[24] In 2018, the Directive was replaced by the GDPR.[25] The GDPR made some significant differences to the data protection landscape in the EU. As a regulation, rather than a directive, the GDPR had to be implemented as defined by the EU across member states.[26] There was no member-state option for flexible implementation, which was important in relation to information privacy because some member states, most notably Ireland and the UK, had quite different regimes with regard to continental Europe. The GDPR thus provided a concerted attempt to harmonise the application of information privacy law across the EU.

[22] Charter of Fundamental Rights of the European Union OJ [2010] C 83/02.
[23] Article 8(1) states, 'Everyone has the right to the protection of personal data concerning him or her.'
[24] Schwartz and Peifer.
[25] Regulation (EU) 2016/679 of the European Parliament and of the Council of 27 April 2016 on the protection of natural persons with regard to the processing of personal data and on the free movement of such data, and repealing Directive 95/46/EC (General Data Protection Regulation).
[26] Yves Poullet, 'Is the General Data Protection Regulation the Solution?' (2018) 34 Computer Law & Security Review 773.

The GDPR also introduced an extra-territorial effect. Under Article 3(1), it could apply to non-EU companies that are outside of the EU, where those companies (a) offer goods and services in the EU, whether free or paid,[27] or (b) monitor the behaviour of individuals in the EU.[28] The GDPR also introduced a range of new micro-right[29] protections, such as the new data portability right under Article 20 and the codification of a right to be forgotten, as a form of mandated erasure of personal information, in Article 17. It also enhanced existing protections under the Directive that were largely underutilised – most notably a controversial right to explanation for solely automated decision making across Articles 13, 14 and 15[30] and a right of objection to solely automated processing under Article 22. The GDPR also increased the scope for financial penalties. Nation-state regulators can work together across the EU, and the maximum fine now applicable is 20 million euros or 4 per cent of a company's global turnover.[31] Despite all these enhancements, it is also important to recognise that the GDPR is still largely based on the same core principles as detailed in the Directive.[32]

The third and final founding instrument is the OECD's *Guidelines on the Protection of Privacy and Transborder Flows of Personal Data*,[33] which crystallised transnational improvements in 1980 and set the foundation for legislative developments in a number of OECD member countries.[34] The OECD contended that the 1970s was an intensive period of legislative investigation and activity about the protection of privacy with respect to the collection and use of personal information. Member countries of the OECD had a common interest in the protection of individual privacy and in the reconciliation of fundamental and competing values involved in automatic data processing and transborder flows of personal information. The Guidelines were therefore intended as a consensus-based guide from which national members could structure their own information privacy legislation.

---

27 The examples used at Article 3(1) include 'a website or online service, such as cloud storage, that can be used by EU citizens'.

28 For example, 'a social network that has EU members or allows EU members to join'.

29 Orla Lynskey, 'Aligning Data Protection Rights with Competition Law Remedies? The GDPR Right to Data Portability' (2017) 42 European Law Review 793, 809.

30 Sandra Wachter, Brent Mittelstadt and Luciano Floridi, 'Why a Right to Explanation of Automated Decision-Making Does Not Exist in the General Data Protection Regulation' (2017) 7 International Data Privacy Law 76.

31 Article 83(5), GDPR.

32 Tal Z Zarsky, 'Incompatible: The GDPR in the Age of Big Data' (2017) 47 Seton Hall Law Review 1020, 1004.

33 OECD, *Guidelines on the Protection of Privacy and Transborder Flows of Personal Data* (OECD 1980).

34 Michael Kirby, 'Twenty-five Years of Evolving Information Privacy Law: Where Have We Come from and Where Are We Going?' (2003) 21 Prometheus 467.

As with the HEW report, and the Council of Europe Convention 108, the OECD Guidelines were concerned with the maintenance of balance. On this occasion, the balance was the harmonisation of different legislation to protect privacy and preserve the integrity of transborder flows of personal information, with a focus on information privacy.[35] The Guidelines were therefore an attempt to reduce the restrictions that inhibited the transfer of personal information and to strengthen the free information flow between member countries. The notion of balancing competing claims is consequently at the heart of the Guidelines as it attempts to consider both individual and organisational interests through a range of principled measures similar in idea to other jurisdictions. The Guidelines provided eight core principles of data collection, storage and use for application by member countries.

1. *Collection limitation:* guarantees that the collection of personal data is within lawful and fair means and where appropriate is conducted with the knowledge and consent of the individual
2. *Data quality:* requires data collectors to collect personal data for relevant purposes only and to ensure that collected data is accurate, complete and up to date
3. *Purpose specification:* states that the purpose for which personal data is to be used must be stated at the time of collection and that subsequent use must be limited to that purpose, unless individuals are notified of additional uses before that reuse takes place
4. *Use limitation:* states that personal data should only be disclosed or used in accordance with the consent of the individual or by authority of law
5. *Security safeguard:* requires that personal data must be kept in reasonably secure conditions
6. *Openness:* states that organisations should implement a general policy of openness about data collection developments, practices and policies
7. *Individual participation:* confirms that an individual should retain certain rights over the collection, storage and use of their information
8. *Accountability:* confirms that a data-collecting organisation should be accountable for complying with these principles

The original Guidelines were updated by the OECD in 2013 in recognition of the technological advances that had taken place in the previous thirty years.[36] It was particularly noted that the volume

---

[35] Fred H Cate, 'The EU Data Protection Directive, Information Privacy, and the Public Interest' [1994] Iowa Law Review 431.
[36] OECD, *The OECD Privacy Framework* (OECD Publishing 2013) 19.

of personal data collections – which had concomitant risks to individual privacy – had increased markedly during this period. Most notably

[p]ersonal data is increasingly used in ways not anticipated at the time of collection. Almost every human activity leaves behind some form of digital data trail, rendering it increasingly easy to monitor individuals' behaviour. Personal data security breaches are common. These increased risks signal the need for more effective safeguards in order to protect privacy.[37]

Despite the radical technological changes that had occurred since the 1980s, the expert group established to examine the Guidelines' current applicability decided that original principles still provided the appropriate level of balanced application necessary to ensure adequate protections. The 2013 revised Guidelines were therefore an update rather than a fundamental rethink of the original core principles.[38] Recommendations focused principally on the provision of additional management processes to supplement the administration of existing principles and to incorporate new developments such as mandatory data breach notification.[39] Perhaps unsurprisingly, given the limited ambitions of the 2013 revision, the new Guidelines have not had the same impact regarding the promulgation of new legislation or the updating of existing legislation as its policy predecessor.

The effect of the original Guidelines, as noted further on, was significant, as it spurred several national information privacy laws.[40] Take, for example, Australia. The first Australian privacy laws started to appear in the early 1970s and focused on the use of technology for eavesdropping or wiretapping.[41] These laws were driven by specific technological advancements and were not universally adopted throughout Australia. Nevertheless, concerns regarding the protection of individual privacy culminated in the mid-1970s when the Australian Law Reform Commission (the ALRC) was asked to undertake a wide-ranging review of privacy in Australia, including issues relating to the collection, storage and use of personal information.[42] The ALRC reported back to the Australian government in 1983 and made recommendations

---

[37] Ibid 20.
[38] Ibid 22.
[39] Ibid.
[40] Ibid 77–9.
[41] For example, the Invasion of Privacy Act 1971 (Qld), which was enacted following concerns relating to the use of listening devices for surveillance purposes. See also Listening & Surveillance Devices Act 1972 (SA).
[42] Australian Law Reform Commission, 'Privacy' (ALRC 22, Australian Law Reform Commission, 1983).

regarding the handling of personal information, largely based on the OECD's Guidelines.[43] On 1 January 1984, the Hawke government, with a degree of symbolism, agreed to accord with the OECD Guidelines. However, it took four years before the development of legislation was finally implemented, and the road to implementation was mired by political controversies and contention.[44]

The Privacy Act was eventually passed by the Commonwealth Parliament to give effect to an agreement to implement the OECD Guidelines, as well as Article 17 of the International Covenant on Civil and Political Rights.[45] That said, the focus of the Privacy Act is very much on the former rather than the latter. As Graham Greenleaf has noted,

[i]n Australia privacy is politics because the constitutional and other institutional protections of privacy are so weak.[46]

The Act initially regulated the conduct of agencies, in a public sector context, including Australian government agencies and the government of the Australian Capital Territory (ACT). Private sector organisations were later included, some twelve years after the implementation of the Privacy Act, when the Privacy Amendment (Private Sector) Act 2000 (Cth) extended the scope of the Act to certain private sector organisations. The basic purpose of the Act remains the same both for government agencies and private sector organisations: the Act regulates how those bodies are permitted to handle and manage personal information. The OECD Guidelines, and their implementation through the Privacy Act, also form the basis of separate state-based legislation that regulates certain state government agencies.[47] As such, the various federal and state information privacy laws share the same foundations both in terms of content and largely in application.

The Act does not prescribe privacy rights to individuals per se but rather seeks to implement a principled approach to privacy regulation that is technology neutral and not corrosive of other rights and freedoms

---

[43] The then president of the ALRC, Michael Kirby, was also the chair of the OECD expert group that developed the 1980 Guidelines.

[44] For a detailed overview, see G W Greenleaf, 'Privacy in Australia' in G W Greenleaf and James B Rule (eds), Global Privacy Protection: The First Generation (Edward Elgar Publishing Limited 2008).

[45] Article 17 states, 'No one shall be subjected to arbitrary or unlawful interference with his privacy, family, home or correspondence, nor to unlawful attacks on his honour and reputation.'

[46] Greenleaf 172–3.

[47] Mark Burdon and Alissa McKillop, 'The Google Street View Wi-Fi Scandal and Its Repercussions for Privacy Regulation' (2013) 39 Monash University Law Review 702.

enjoyed in a liberal democracy.[48] The Privacy Act initially had two sets of privacy principles. The Information Privacy Principles (IPPs) applied to Commonwealth and ACT government agencies. The IPPs form the basis of separate state-based legislation that also regulates state government agencies as highlighted later in this discussion.[49] The National Privacy Principles applied to private sector organisations. In 2014, following another major review of the Privacy Act by the ALRC in 2008, the two were merged into the Australian Privacy Principles (APPs) that currently apply, albeit differently to public and private sector data collectors. Finally, it is also important to note that there has been little judicial interpretation of the Privacy Act's key components, and the interpretation of many of the Act's central provisions are still a matter of some speculation,[50] including the definition of 'personal information', as we will go on to discuss.

6.2.2   Jurisdictional Approaches

The HEW report, the Council of Europe Convention and the OECD Guidelines have been at the forefront of the development of what have been termed 'first generation information privacy laws'.[51] There are obvious similarities between the three instruments that information privacy laws generally reflect.[52] All of these laws have organisational-oriented controls founded on privacy principles or fair information practices developed before the advent of sensorisation and data analytics. However, while there are similarities, there are also some significant differences, particularly regarding the need for individual protections at the point of collection, as we will highlight.

These differences show that the implementation of information privacy laws have taken essentially different tracks even though their genesis resorts from similar conceptual foundations. That is not surprising. As we have already seen, the concept of information privacy is essentially contested, and thus the interpretation of what weight of an individual's right to control their personal information is in competition with other social rights and interests.[53] The application of information privacy

---

[48] Australian Law Reform Commission, *For Your Information: Australian Privacy Law and Practice* (Law Reform Commission 2008) 233–48.

[49] Greenleaf 152.

[50] Ibid.

[51] Jonathan Zittrain, 'Privacy 2.0' [2008] The University of Chicago Legal Forum 65.

[52] Marc Rotenberg, 'Fair Information Practices and the Architecture of Privacy' Stanford Technology Law Review <http://stlr.stanford.edu/STLR/Articles/01_STLR_1> 2.

[53] Bennett and Raab 13.

legal regimes is subject to discussion amongst different legislative jurisdictions contesting scope and application.[54] Information privacy laws
are manifestations of political processes which have implications for the
implementable scope of such laws. Jurisdictional information privacy
laws consequently reflect the wider social, legal and policy values of
individual jurisdictions.[55] United States attitude towards information
privacy law reflects this point.

The sectoral approach[56] to information privacy in the United States
has been characterised as 'sporadic'[57] and 'reactive'.[58] The regulatory
focus of US information privacy law is the general curtailment of government powers in combination with laws that govern industry-specific
practices and various types of sensitive information. The existence or
nonexistence of information privacy regulation at the federal level is
specific to circumstances or sectors. For example, as noted earlier,
the US Privacy Act[59] provides a range of FIPPs that US government
agencies must comply with regarding the handling of personal information. However, other federal information privacy laws have different remits. The Gramm Leach Bliley Act (GLBA)[60] creates privacy
protections for personal financial information within the specific remit
of the financial services sector. The Health Insurance Portability and
Accountability Act (HIPAA)[61] consigns legal protections in relation to
identifiable health information held in the medical and health insurance
sectors. In a different vein, the Children's Online Privacy Protection
Act (COPPA)[62] governs restrictions on the collection of online personal
information from children under the age of thirteen.

Alongside these sector-based laws, there is a collection of other laws
that have been developed to provide a remedy for specific issues which
have become sufficiently politicised to warrant legislative action.[63] For
example, the Drivers Privacy Protection Act (DPPA)[64] was enacted to
restrict the disclosure of driver license information by state authorities

---

[54] Regan, *Legislating Privacy* 16.
[55] Bennett and Raab 125.
[56] Gellman 195 describing sectoral as 'no general privacy laws, just specific laws cover-
      ing specific records or record keepers'.
[57] Ibid describing the legal structure for US privacy protection as a 'patchwork quilt'.
[58] Bennett and Raab 37 regarding reactivity as a weakness of sectoral regimes.
[59] The Privacy Act of 1974, 5 U.S.C. § 552a.
[60] Financial Services Modernization Act of 1999 15 U.S.C. §§ 6801–6809 (1999).
[61] Health Insurance Portability and Accountability Act of 1996 45 C.F.R. §§ 160, 162
      and 164 (1996).
[62] Children's Online Privacy Protection Act 15 U.S.C. § 6501–6506 (1998).
[63] Bennett and Raab 37.
[64] Drivers Privacy Protection Act of 1994 18 U.S.C. § 2725 (1994).

following the murder of actress Rebecca Schaeffer, in which an assailant used publicly available driver license information to stalk and then murder Ms Schaeffer.[65] The requirements of the DPPA have also been instrumental in restricting the sale of driver license information by state agencies to commercial entities. The Video Privacy Protection Act[66] was enacted following a controversy involving Supreme Court nominee Robert Bork, when details of his video-watching habits were gained by the media.[67]

The myriad of information privacy legislation has also been replicated at state level and city level also.[68] Some states have implemented laws to provide general statutory rights of privacy that are akin to tort law protections and thus govern areas such as common law invasions of privacy.[69] Other states, like their federal counterparts, have also enacted a number of sectoral-based laws aimed at certain industry practices. For example, in addition to federal laws, some states have developed legislation specifically covering the use of personal information in relation to certain information – such as video rental records, as just referred to.[70] Schwartz contends that a duopoly exists between federal and state laws, in which federal laws deliver specified benchmarks that allow state laws further room for experimental development.[71]

On the other hand, the EU and OECD countries, like Australia, through comprehensive legal frameworks, adopt a different approach to sectoral regimes. They establish information privacy rights for individuals and define obligations for data-collecting organisations regardless of industrial sector. Enforcement mechanisms operated by comprehensive information privacy regimes are also different to those found in sectoral regimes. Most comprehensive frameworks employ specific supervisory authorities with given sets of legislative powers to protect the rights of individuals and impose compliance obligations upon organisations – a necessary condition of an effective information privacy regime.[72]

---

[65] Solove, *Understanding Privacy* 69; Regan, *Legislating Privacy* 207 regarding the use of state driver license information to harass pregnant mothers who visited abortion clinics.

[66] The Video Privacy Protection Act of 1998, 18 U.S.C. § 2710.

[67] See Paul M Schwartz, 'Preemption and Privacy' (2009) 118 Yale Law Journal 902, 935–6 providing a comprehensive overview to the development of the law including details of congressional outrage.

[68] Ira Rubinstein, 'Privacy Localism' (2018) 93 Washington Law Review 1961.

[69] Joel R Reidenberg, 'Privacy in the Information Economy: A Fortress or Frontier for Individual Rights?' (1992) 44 Federal Communications Law Journal 195, 228.

[70] Patricia L Bellia, 'Federalization in Information Privacy Law' (2009) 118 Yale Law Journal 868.

[71] Schwartz, 'Preemption and Privacy' 919.

[72] Bennett and Raab 113.

This brief history of information privacy law's development highlights that legal protections are linked to broader social policy considerations and political outcomes. As noted in Chapter 5, many authors have argued that conceptually the constituent elements of information privacy law do not match the inherent complexities of social life, including the intrinsic relationships involved in personal information exchange processes. It is argued that information privacy laws have some inbuilt weaknesses about how the roles of individuals and data-collecting institutions are constructed.

These legal constructions 'lay down the tramlines' about how organisations understand their legal obligations and individuals understand themselves in relation to these organisations.[73] Information privacy laws and legal regimes are too heavily institutionalised and focus too much upon the regulation of typified institutions rather than the governance of information relationships.[74] These underlying criticisms of information privacy will now be examined according to the type of information regulated in different jurisdictions. It becomes clear that different jurisdictions place diverse social and political priorities regarding the categorisation of personal information – a point that becomes important in Chapter 7's legal analysis of smart home insurance business models that utilise sensorised data collections.

6.3   Regulated Information: PII, Personal Data and Personal Information

Information privacy law is predicated on the regulation of specified types of information that are intrinsic to individual data exchange processes. Section 6.2 highlighted the information privacy law differences between jurisdictions that emanate from similar principled roots. Not surprisingly, the same can be said for the type of information that is regulated. The sectoral US approach regulates defined types of 'personally identifiable information' that are designed to meet a specific statutory purpose.[75] As a result, there are different definitions of PII across a range of statutes. Comprehensive frameworks such as those of the EU and Australia have universal notions of the type of information that is covered by information privacy laws – typically defined as 'personal data' in the GDPR or 'personal information' in the Australian Privacy Act.

---

[73] Ibid 35.
[74] Ibid.
[75] Birnhack 264.

The definitional scope of what data is classified as personal information is crucial to resolve, as information privacy law will only apply to specified types of regulated information.[76] The classificatory basis of regulated information is therefore important because it belies many of the political considerations inherent to the application of information privacy law. The US situation, in comparison to those of the EU and Australia, is an important case in point. Each jurisdiction has a different method of classification, which has an impact on the scope of application, as detailed in Chapter 7 regarding smart home sensor data. In the meantime, the conceptual construction of each type of information is detailed in this section.

6.3.1   Conceptual Differences

As noted previously, the sectoral approach of the US framework is fundamentally different to the EU and the OECD tracks. Rather than having a comprehensive information privacy law that applies to all types of personal information collections, regulatory coverage is dependent upon whether data collection activities are covered by a specific piece of sectoral legislation. There is not a uniform definition of PII in the US system, given that each individual construction is dependent upon ameliorating a specifically identified statutory harm.[77] To complicate matters further, nor is there a universal method of classifying PII, because its conceptual roots emerge from Warren and Brandeis' property-analogous application of reputational protection as well as information privacy.[78] The US notion of PII therefore engenders a much greater focus on property protections as the primary basis of construction than found in other jurisdictions.[79]

The result of these regulatory and conceptual entanglements is a panoply of different PII definitions with no uniform understanding of how to conceptualise classifications of data for regulation in information privacy law frameworks. Take for example one of the first US information privacy laws: as noted earlier, the Privacy Act. The Act's classificatory trigger does not regard PII per se. Rather, it regards information retrieval from systems of records when the retrieval is undertaken with

---

[76] Daniel Solove and Paul M Schwartz, 'The PII Problem: Privacy and a New Concept of Personally Identifiable Information' (2011) 86 New York University Law Review 1814, 1816.

[77] Paul M Schwartz and Daniel J Solove, 'Reconciling Personal Information in the United States and European Union' (2014) 102 California Law Review 877.

[78] Strandburg 6.

[79] Purtova, 'The Illusion of Personal Data as No One's Property'.

an individual's name or identifying number or other identifying particular assigned to the individual.[80] A narrow information privacy protection is applied relating to specified PII identification processes for a specified information retrieval process. As such, the Act only covers computer searches that identify an individual via a specific form of identifier.[81] The Privacy Act is fundamentally different to other first-generation information privacy law because it does not establish obligations for data collectors at the point of personal information. The Cable Communications Act 1984 was the first statute that established the generalised approach that PII collections triggered legislative obligations based around fair information privacy practices.[82] Other sector- or information-specific statutes followed, each with their own distinct definition of PII.

Schwartz and Solove highlight the weaknesses of the US approach to PII and attempt to provide conceptual insight regarding classification logics via a three-element typology that distinguishes between different approaches to PII across a range of statutes. The Tautological Approach is the broadest and simply defines PII as 'any information that identifies a person', such as the definition found in the Video Privacy Protection Act (VPPA).[83] The breadth of the definition's scope is both a strength and a weakness. As an open-textured standard, it can evolve over time to encompass a range of different data that is adaptable to changing technological circumstances. However, the definition does not assist in determining PII from non-PII, as it simply states that PII is PII.[84] The Non-Public Approach differentiates publicly available information that can identify from non-publicly available information, such as evidenced by the GLBA.

The GLBA aims to protect consumer financial privacy by placing limits on when a financial institution can disclose a consumer's 'non-public personal information' (NPI) to third parties. NPI is any 'personally identifiable financial information' that a financial institution collects about an individual in connection with providing a financial product or service, except when that information is otherwise publicly available. Types of NPI typically include information provided by individuals in relation to a financial product or service. For example, it can be identifying data supplied at registration or application; transactional information regarding use of a service or product, such as account

[80] Solove and Schwartz 1823.
[81] Ibid 1824.
[82] Ibid 1817.
[83] Ibid 1829. The obligations of the VPPA for video tape service providers are triggered when PII is linked to the purchase, requesting or obtaining of video material.
[84] Ibid.

numbers, payment histories or credit purchases; or any other information a financial institution receives about an individual in connection with providing the financial product or service (e.g. a credit-matching score).[85] The Non-Public Approach effectively severs the process of identifying in relation to publicly available data. Consequently, 'the private status of data often does not match up to whether it can identify a person or not'.[86]

Finally, the Specific-Types Approach regards rule-based amelioration of a specific type of harm, such as data breach notification. Forms of US data breach notification law attempt to mitigate the specific harm of identity theft, and they do so by regulating specified forms of personal information in combination with other information.[87] These typically involve primary forms of identification, such as name, with secondary forms such as social security number, drivers licence, biometric information or passport details.[88] COPPA also contains a similar definition, in which 'personal information' is individually identifiable information about an individual collected online that includes, amongst other things, name, address and email address.[89]

A criticism of this typology is that it narrowly corrals information privacy issues into a specified issue and may regularly require updating to keep up with technological change. An example relates to the Federal Trade Commission's (FTC) COPPA rules, which added an additional form of specified information, a persistent identifier that relates to a customer number or serial number that can be associated with individually identifying information.[90] Additional and further rule making is therefore a prerequisite requirement of the Specific Type Approach.[91]

Taking on board the foregoing discussion, it is perhaps not surprising that the US sectoral framework has been labelled a patchwork of inconsistencies.[92] Information privacy law, as the actioning of a control-based concept, only provides control procedures for information that it regulates.

---

[85] 'How To Comply with the Privacy of Consumer Financial Information Rule of the Gramm-Leach-Bliley Act' (Federal Trade Commission, 2002) <www.ftc.gov/tips-advice/business-center/guidance/how-comply-privacy-consumer-financial-information-rule-gramm#obligations> accessed 20 June 2019.

[86] Solove and Schwartz 1830.

[87] Paul M Schwartz and Edward J Janger, 'Notification of Data Security Breaches' (2007) 105 Michigan Law Review 913.

[88] Mark Burdon, 'Contextualizing the Tensions and Weaknesses of Data Breach Notification and Information Privacy Law' (2010) 27 Santa Clara Computer and High Technology Law Journal 63, 107.

[89] Solove and Schwartz 1831.

[90] Ibid 1832.

[91] Ibid 1835.

[92] Schwartz and Peifer 136.

The ability and effectiveness of any information privacy law is thus intrinsically linked to the information it seeks to protect. Solove and Schwartz note that

[i]f PII is defined too narrowly, then it will fail to protect privacy considering modern technologies involving data mining and behavioral marketing. Technology will thus make privacy law irrelevant and obsolete. On the other hand, if PII is defined too broadly, then it could encompass too much information, and threaten to transform privacy law into a cumbersome and unworkable regulation of nearly all information. Privacy law must have coherent boundaries, which adequately protect privacy, and which can be flexible and evolving.[93]

The comprehensive regimes of the EU and the OECD-based countries employ very different approaches to the classification of personal data and personal information. Nevertheless, the foregoing quote is relevant to both jurisdictions, as evidenced by the differences arising from the GDPR and the Australian situation, in which the substitution of one word, from 'relating' to 'about', has significant intended effect in coverage and application. The semantic change also signifies different policy emphases about the breadth accorded to information privacy protection in each jurisdiction. That said, unlike in the United States, the conceptual approach to classification in both jurisdictions applies from the same coherent and comprehensive base.

Personal data is defined in the GDPR under Article 4(1) as

any information relating to an identified or identifiable natural person ('data subject'); an identifiable person is one who can be identified, directly or indirectly, in particular by reference to an identifier such as a name, an identification number, location data, online identifier or to one or more factors specific to the physical, physiological, genetic, mental, economic, cultural or social identity of that person.

Similarly, in Australia, personal information is defined under Section 6(1) of the Privacy Act as

information or an opinion about an identified individual, or an individual who is reasonably identifiable: (a) whether the information or opinion is true or not; and (b) whether the information or opinion is recorded in a material form or not.

Three component concepts pervade in both definitions: (1) 'relating to' and 'about', which define the connective process that links an individual to a recognisable and applicable process of identity, (2) the two states of identity that flow from that process – namely, identified and identifiable,

---

[93] Solove and Schwartz 1827.

or reasonably identifiable in the Australian context and (3) additional considerations of sensitivity. There is general agreement about how (1) and (2), namely, the different states of identity are conceptualised in both jurisdictions which entail context-independent and context-dependent approaches. Both jurisdictions also recognise that different forms of personal information can have greater levels of sensitivity attached to them and have separate definitions of 'sensitive personal data' and 'sensitive information' which require higher forms of regulatory obligation in handling. Both types include specific categories of personal information that give rise to additional sensitivities regarding use – such as racial origins, political or philosophical beliefs or sexual orientation. Sensitive data and information also include biometric data that can be used for identification purposes or for biometric verification.

Both definitions of personal data and personal information are predicated on the ability to identify an individual from certain information, and this can be done in two ways, where the following happens to an individual's identity.

1. *It is identified.* A piece of data can identify an individual – for example, where the data is a name or photograph of an individual. Identified data is more redolent to the types of US PII detailed earlier.
2. *It is identifiable or reasonably identifiable.* A single piece of data is not able to identify an individual, but that data can be used to cross-reference with other information that then identifies an individual. For example, a home address does not automatically reveal identification, but it can be used as a central point to cross-reference other information which then reveals identity.[94] Note also the different identifiable process requirements in the GDPR and the Privacy Act. Under the former, an identifiable person is one who can be directly or indirectly identified by cross-referencing an identifier or identity factors. In the latter, the identifiable process requisites are not necessarily restricted to cross-referencing by an identifier or other identity factors, but the process of identifiability must be reasonable.[95]

Both definitions therefore have a broad application which features both context-dependent and context-independent approaches to the identification of personal information.[96] A *context-independent approach* enables

---

[94] Australian Law Reform Commission 309.
[95] Note also the use of 'reasonably likely' in Recital 26 of the GDPR.
[96] Mark Burdon and Paul Telford, 'The Conceptual Basis of Personal Information in Australian Privacy Law' (2010) 17 Murdoch Elaw Journal 1. For clarification of both distinctions, see Sharon Booth and others, *What Are 'Personal Data'? A Study Conducted for the UK Information Commissioner* (2004).

the categorisation of personal information without recourse to the social context within which the information is used. It represents the 'identified' state of both definitions. The removal of social context simplifies the categorisation of personal information, because it is possible to make a definitive prediction of what information is always likely to be classified as personal information. For example, a name is always likely to reveal identification, so therefore a name will always be personal information.

Alternatively, a *context-dependent approach*, as defined by the 'identifiable' or 'reasonably identifiable' state, deems that personal information can only be identified by examining the social context within which a piece of information is used. This makes definitive prediction virtually impossible, because all information could be classed as personal information in the right circumstances, which are likely to be inherently subjective. For example, as mentioned earlier, a home address does not automatically reveal identity, but it can in certain circumstances, by cross-referencing it with other information.

Under both definitions, data does not have to identify a person directly for it to be classified as personal data or information. For example, it is possible for data to be classed as personal if a person is not mentioned by name but can nonetheless be identified. However, information that simply allows an individual to be contacted, such as a telephone number or address, would not in itself be deemed personal information, as the GDPR, and particularly the Privacy Act, was not intended to provide an unqualified right to be left alone.[97]

As noted earlier, however, once other information accretes around a specific piece of information and a data collector is then able to target that individual 'by linking data in an address database with particular names in the same or another database, that information is personal information'.[98] At this point, both the GDPR and the Privacy Act recognise that the character of the information set as a whole tilts toward being personal. From an information privacy perspective, certain pieces of information such as home or IP addresses can act as an identifier to link different datasets together.[99] Linking datasets increases the likelihood that the identity of the subject is identifiable from the set. The status of information as personal data or personal information therefore

---

[97] Australian Law Reform Commission 309.
[98] Ibid.
[99] Normann Witzleb and Julian Wagner, 'When Is Personal Data about or Relating to an Individual? A Comparison of Australian, Canadian, and EU Data Protection and Privacy Laws' (2018) 4 Canadian Journal of Comparative and Contemporary Law 293.

has an important element of context – namely, the context and the interrelationship of each of the available information components and the extent to which they collectively make identification possible. The context element, in the form of digitally generated metadata, is specifically recognised in the GDPR. Recital 30 states as follows:

Natural persons may be associated with online identifiers provided by their devices, applications, tools and protocols, such as internet protocol addresses, cookie identifiers or other identifiers such as radio frequency identification tags. This may leave traces which, when combined with unique identifiers and other information received by the servers, may be used to create profiles of the natural persons and identify them.

Both jurisdictions also distinguish between actual and theoretical forms of identification as a determining factor for deciding whether a piece of information could be categorised as personal.[100] An individual can be identifiable when they can be identified from information held in the possession of an organisation and access to that information would not be prohibitive in terms of cost or difficulty.[101] For example, the test for confirming whether a piece of information is likely to be personal information, therefore requires 'a consideration of the cost, difficulty, practicality and likelihood that the information will be linked in such a way as to identify [an individual]'.[102] As a limiting factor, the ALRC also rejected the idea that the test should include whether an individual is potentially identifiable.[103] The Article 29 Working Party has also decided similarly.[104]

The context of the information-collecting organisation is consequently important. A piece of information is less likely to be categorised as personal if the collecting organisation has little information resources and would not be able to easily or quickly identify an individual by cross-referencing the information in question with other information

---

[100] Australian Law Reform Commission 307, where the Commission states, 'While it may be technically possible for an agency or organisation to identify individuals from information it holds, for example, by linking the information with information held by another agency or related organisation, it may be that it is not practically possible. For example, logistics or legislation may prevent such linkage. In these circumstances, individuals are not "reasonably identifiable".'

[101] Office of the Australian Information Commissioner, *What Is Personal Information?* (OAIC 2017).

[102] Australian Law Reform Commission 308.

[103] Ibid, because a 'great deal of information is about potentially identifiable individuals but ... identifying the individuals would involve unreasonable expense or difficulty, and is unlikely to happen'.

[104] Article 29 Data Protection Working Party, Opinion 4/2007 on the Concept of Personal Data (01248/07/EN WP 136, 2007) 15.

held by it. However, the corollary also occurs. It is more likely that a piece of information will be categorised as personal information if the collecting organisation has significant information resources and the cross-referencing is not prohibitive in terms of cost or difficulty.[105]

6.3.2 Judicial Considerations

Thus far, the conceptual basis of identified and identifiable states is similar in both jurisdictions. However, the effect of different connective elements, in the form of 'relates' and 'about', has led to significant differences in application, particularly involving the complex issue of whether certain types of mobile device metadata, including IP addresses, would be classified as personal in both jurisdictions. In turn, this seemingly small and inconsequential difference sheds light on the very different social policy ambitions of the EU and Australian information privacy law frameworks.

The Article 29 Working Party is the EU's data protection policy interpreter and instigator. The Working Party's opinions are persuasive rather than binding and refer to the definition of personal data in the previous Data Protection Directive. However, given that the definition has not changed, the same interpretation will be applied under the GDPR. The Article 29 Working Party, in guidance about the concept of personal data, under the Data Protection Directive, clearly indicated that the definition of personal data was to be construed broadly as a means of protecting fundamental rights and freedoms of EU citizens. The definition of personal data was thus intended to be interpreted with 'considerable flexibility'[106] to give effect to a right of privacy regarding data processing.[107] The degree of flexibility is evident in the Working Party's consideration of 'relating to'. The Working Party decided that, in general terms, information relates to an individual when it is about that individual.[108] In that sense, there is an obvious, at face, similarity between the definition of personal data and personal information under the Australian Privacy Act. However, the connective element of what information will be about an individual, in a 'relating to' sense, is constructed broadly across three elements.

---

[105] Consequently, virtually any piece of data could be personal information in the right context, including a black suit. See Bart van der Sloot and Frederick Zuiderveen Borgesius, 'The EU Data Protection Regulation: A New Global Standard for Information Privacy' <https://bartvandersloot.com/onewebmedia/SSRN-id3162987.pdf> citing the Working Party's 2007 Opinion at page 13: '"the man wearing a black suit" may identify someone out of the passers-by standing at a traffic light'.

[106] Article 29 Data Protection Working Party 5.

[107] Ibid 4.

[108] Ibid 9.

The first element is content. Put simply, information is about a person if the content of the information is clearly about that individual – for example, a medical diagnosis that clearly relates to or is directly about a patient. The second element is purpose. Information can be about an individual where its purpose is to 'evaluate, treat in a certain way or influence the status or behaviour of an individual'.[109] For example, call logs of office phones could be classed as personal data. Call log data could be about a person if the data's purpose is to identify specific employee calling habits or monitor employee communications. The third element is result. Data can be about an individual, regardless of whether content or purpose exist, where data is likely to impact on an individual's rights and interests, considering all contextual circumstances.[110] Impact in this regard refers to being treated differently from others because of data processing. The three elements are alternative rather than cumulative and therefore provide a wide scope for when data will be about individuals that goes beyond a pure subject matter characterisation of data.

The effect of breadth can also be seen in the Working Party's guidance on geolocation services and smart mobile devices.[111] The Working Party Opinion provides a detailed overview of the legal implications of collecting geolocation data, including Wi-Fi header data, across a number of different telecommunications sectors that involve a number of actors. Part of the analysis regards what data should be classed as personal data across three types of geolocation infrastructure: GPS, Global System for Mobile Communications (GSM) and Wi-Fi. Importantly, as part of the immediate considerations being discussed here, the Working Party decided that, as a matter of policy principle, identifiers of smart mobile devices, when utilising geolocation services, should be classified as personal data. The inextricable link between smart mobile devices and individuals was always likely to give rise to either direct or indirect identification.[112] Smart mobile identifiers could be combined with relative ease with other location data to single out an individual.[113]

In perhaps a more guarded opinion, relating to IoT, the Working Party further indicated that data generated by wearable devices for health and the home would most likely be classed as personal data.

---

[109] Ibid 10.
[110] Ibid.
[111] Article 29 Data Protection Working Party, Opinion 13/2011 on Geolocation Services on Smart Mobile Devices (881/11/EN WP 185, 2011).
[112] Ibid 9.
[113] Frederik J Zuiderveen Borgesius, 'Singling Out People without Knowing Their Names: Behavioural Targeting, Pseudonymous Data, and the New Data Protection Regulation' (2016) 32 Computer Law & Security Review: The International Journal of Technology Law and Practice 256.

IoT stakeholders aim at offering new applications and services through the collection and the further combination of this data about individuals – whether to measure the user's environment specific data 'only', or to specifically observe and analyse his/her habits. In other words, the IoT usually implies the processing of data that relate to identified or identifiable natural persons, and therefore qualifies as personal data.[114]

The Working Party specifically noted that sensor data, particularly from wearable devices, could be used in conjunction with the device's identifiers to generate a unique digital fingerprint that could be tracked in a proximate location or across locations.[115] It is therefore likely that IoT sensor data would be personal data particularly where such data 'may allow discerning the life pattern of a specific individual or family'.[116]

Equally, the Court of Justice of the European Union (CJEU) has been willing to consider the definition of personal data expansively and has found that both static and dynamic IP address data should be classed as personal data. In *Scarlet Extended*, the court had to determine whether a static IP address should be classed as personal data. The issue before the court involved illegal file sharing and copyright infringement actions against the applicant. Scarlet, an Internet service provider (ISP), refused a legal request by SABAM, a content management company, to install a content-filtering system to monitor and prohibit customer use of certain IP addresses. Scarlet objected on several grounds, including that the installation of a secret filtering system would breach the Data Protection Directive, given that filtering involves the processing of IP addresses, which, Scarlet argued, was personal data.[117]

The court determined, based on the technological analysis supplied to the court, that a static IP address held by an ISP should be classed as personal data, as an individual could be precisely identified from the use of that data.[118] The court then had to determine a balancing of conflicting rights in the context of the EU rights-based framework – namely, the rights of copyright holders to prevent infringement versus the data protection rights of individuals under Articles 8 and 11 of the EU Charter.[119] The broad nature of preventative monitoring became the key point of analysis.

---

[114] Article 29 Data Protection Working Party, Opinion 8/2014 on the Recent Developments on the Internet of Things (14/EN WP 223, 2014) 4.
[115] Ibid 8.
[116] Ibid 10.
[117] *Scarlett Extended SA v Societe belge des autuers, compositeurs et editeurs SCRL (SABAM)*, [26].
[118] Ibid [51].
[119] Ibid [50].

In effect, the filtering system would require the systematic analysis of all user content, including the identification of users' IP addresses 'from which unlawful content on the networks is sent'.[120]

Scarlet involved a clear identification process both from the perspective of static IP address data and an ISP's ability to undertake an identification process. However, the court has also being willing to identify personal data in less clear-cut data and holder situations. In *Breyer v Bundesrepublik Deutschland*, the CJEU had to determine whether the collection of a dynamic IP address, temporarily linked to a host computer and assigned by an ISP, was capable of being classed as personal data when the IP address was collected and held by the federal German government. As noted earlier, to do so, it would have to be shown that the government was able to identify the applicant from the IP address. Such an action would generally be problematic, because usually the ISP was the only organisation that could link the IP address to other data held by it to identify an individual. In a broad reading of the 'reasonably likely' element of identifiable personal data, the court concluded that a dynamic IP address assigned by an ISP could be personal data related to that data being held by the federal government.

The court's decision turned on the wording of Recital 26 in the Directive and the need to take account of data held by a data controller or by 'another person' to identify an individual.[121] It decided that Article 2(a) did not require that all information to enable identification of an individual 'must be in the hands of one person'.[122] This analysis is telling because the court then considered the federal government's capacity to reasonably identify an individual from the dual perspective of it being an online media service and also a competent law enforcement authority that had the ability to obtain dynamic IP address details from any person's ISP. Such actions were commonplace regarding law enforcement investigations of cyberattacks as a basis for initiating legal action. The combination of public accessibility of a government website and the legal means to obtain additional data from an ISP thus became the justificatory nexus for the court's decision.[123]

The interpretation of personal data is thus broad and in keeping with the rights-based nature of that jurisdiction. Australia, as one of the non-EU OECD countries, has a similar conceptual framework for determining personal information, but it is narrower in application, as

120   Ibid [51].
121   Ibid [42].
122   Ibid [43]–[44].
123   Ibid [49].

evidenced by a series of recent litigation. First, though, it is important to explore briefly the transition from the OECD Guidelines' definition of personal data to the one that finds form in the Privacy Act. The Guidelines put forward a similar definition of personal data encapsulated in the Directive and the GDPR. The OECD Guidelines deliberately provided member states with a substantial amount of interpretative leeway, given the differing political considerations accorded to information privacy protection. This interpretive leeway was used significantly by the Australian government for the eventual definition of personal information in the Act.

There is a significant difference between the definition of 'personal data' in the OECD Guidelines and the definition of 'personal information' under Section 6(1). The former regards personal data as relating to 'identified or identifiable individuals', whereas the latter relates to information 'about' an individual in an identified or reasonably identifiable sense. The latter definition of personal information is therefore more constricted in its application because the removal of 'relates to' narrows the situations in which information can identify an individual. The Privacy Act's definition of personal information was also considered by the Australian Law Reform Commission (ALRC) in 2008 as part of its For Your Information Report. The ALRC concluded that personal information should still be information about an individual rather than information that relates to an individual.[124] The principle reason for the use of 'about' as opposed to 'relates to' appears to be consistent with the APEC Privacy Framework.

As noted earlier, the Privacy Act is woefully under-litigated, so many of its key terms have yet to be considered judicially.[125] The definition of personal information, and particularly the meaning of 'about', has recently been considered in a series of decisions involving the same case. However, these decisions have complicated the categorisation of personal information in Australia, which culminated in the Full Federal Court decision of *Privacy Commissioner v Telstra*.[126] Following that decision, it is now necessary to consider data as personal information by considering whether (a) the information is about an individual and if so whether (b) the individual is identifiable from the information. The litigation trail, including the Privacy Commissioner's subsequent guideline, is therefore important.

---

[124] Australian Law Reform Commission 306.
[125] Mark Burdon and Alissa McKillop, 'The Google Street View Wi-Fi Scandal and Its Repercussions for Privacy Regulation' (2013) 39 Monash University Law Review 702, 737.
[126] *The Privacy Commissioner v Telstra Corporation Limited* [2017] FCAFC 4 (19 January 2017).

The Federal Court was asked by the Privacy Commissioner to consider the statutory relevance of 'about'[127] in the previous definition of personal information.[128] The case involved a protracted dispute between journalist Ben Grubb and his mobile service provider, Telstra, about accessing mobile metadata. Telstra provided some data but declined Grubb's access request in relation to location-based cell-tower data. Even though it regularly provided such data for law enforcement agencies upon request, Telstra argued that it was unreasonable for an individual access request because it was too complicated, time-consuming and expensive.[129]

Grubb complained to the Privacy Commissioner, and the commissioner decided that the location metadata was personal information because Grubb could be identified in a reasonably ascertainable sense.[130] The commissioner's decision relied heavily on a Victorian tribunal decision, *WL v La Trobe* (*WL*),[131] which indicated the practical boundaries of reasonably ascertainable categorisation.[132] Both the commissioner's decision and *WL* are predicated on a certain logic regarding the identification of data that is personal information. If an individual's identity is apparent or reasonably ascertainable, or under the current definition identifiable or reasonably identifiable, then data should be deemed to be about an individual and be classed as personal information. 'About', in this sense, gets its statutory application from the notion of identity, which at the time was generally settled law based on *WL*.[133]

That categorisation process was then turned upside down on appeal by Telstra to the Administrative Appeals Tribunal (AAT) regarding the Privacy Commissioner's decision.[134] Deputy President Forgie determined the threshold question to be whether the cell-tower data was about an individual and entailed the 'subject matter' or the character of the information.[135] It was now necessary to determine that the subject matter of any given piece of information is 'about' an individual. If it is not, then 'that is the end of the matter'[136] and it does not matter

---

[127] Ibid [5] per Kenny and Edelman JJ.
[128] The key part of the previous definition being 'apparent or reasonably ascertainable'.
[129] *Ben Grubb and Telstra Corporation Limited* [2015] AICmr 35 (1 May 2015), [86].
[130] Ibid [93]–[102].
[131] Ibid [56]–[57], [70]–[71].
[132] *WL v La Trobe University (General)* [2005] VCAT 2592 (8 December 2005) [42] per Coghlan DP.
[133] See *Ben Grubb and Telstra Corporation Limited* [2015] AICmr 35 (1 May 2015), [81]–[83].
[134] *Telstra Corporation Limited and Privacy Commissioner* [2015] AATA 991 (18 December 2015).
[135] Ibid [98].
[136] Ibid [95].

whether identification can be determined from that information, either in an apparent or a reasonably ascertainable sense.[137]

The AAT decision set the scene for the Privacy Commissioner's appeal to the Federal Court. It should be noted that, rather bizarrely, the commissioner's ground of appeal was not the basis of his original decision – namely, that the cell-tower metadata was personal information. Instead, the issue put forward for adjudication was whether 'about' in the definition of personal information had substantive application.[138] The confused court[139] could therefore only adjudicate on the very narrow question of whether 'about' has statutory purpose, rather than the more pressing question of when metadata can be constituted to be personal information in a reasonably ascertainable sense. Not surprisingly, the court held that 'about' had substantive application and partially affirmed the logic of Forgie DP.[140]

The Federal Court also broadened the scope of the AAT decision by accepting that information can have 'multiple subject matters' and thus 'an evaluative conclusion' is required that considers information in its totality.[141] Even though one piece of information may not be about an individual, it can become so when aggregated with other information. However, the court did not indicate when, and in what circumstances, information can have multiple subject matters, and more importantly, when a multiple subject matter characterisation can be about an individual in the context of identification under the old definition of Section 6(1).

The situation is further confused by the Privacy Commissioner's guideline on categorising personal information. The *Grubb* decisions, highlighted earlier, involved the old definition of personal information.[142] As a result of the confusion caused by the Privacy Commissioner's loss at the Federal Court, a new guideline was issued regarding classification of personal information in relation to the updated definition. The guideline responds to the AAT and Federal Court decisions particularly regarding multiple subject matters. However, the guideline also seems to indicate that even if information is not about an individual, it can nevertheless still be personal information, if it reveals or conveys something about an individual.

---

[137] Ibid [97].
[138] *The Privacy Commissioner v Telstra Corporation Limited* [2017] FCAFC 4 (19 January 2017) [5].
[139] Ibid [80].
[140] Ibid [63], [65].
[141] Ibid [63].
[142] The old definition is information about an individual that is apparent or reasonably ascertainable. Both are intended to have the same statutory application, and the change of definition was to bring the Privacy Act in line with other relevant jurisdictions. See Australian Law Reform Commission 299.

The use of 'about' as the basis for personal information classification has an important limiting effect. It reduces the three elements of identification in the GDPR – namely, content, purpose and result – to focus more attentively on content only. Hence the pure subject matter characterisation adopted in the AAT decision and the ameliorated characterisation of 'multiple subject matters' in the Federal Court that extends content into the possibilities of purpose. Note, however, that the subject matter characterisation of 'about' still needs to be resolved before proceeding to identification, so the test primarily focuses on the content of data.

It is evident that the three jurisdictions we have been discussing differ in their approach to the classification of personal information. The US sectoral approach is fundamentally different to the comprehensive regimes of the EU and Australia. However, it is also clear that there are intended policy differences between the EU and Australia that relate to underlying political judgements about the strength and scope of designated information privacy law regimes. We now look at how information privacy law provides individual protections through the application of a principled approach to personal information exchange. In doing so, we again encounter some foundational differences between the United States and the other two jurisdictions and further differences of degree between the EU and Australia. Given that the focus of the book is the onset of the collected world, Section 6.4 now focuses specifically on the controversial issue of information privacy protections at the point of collection.

6.4   Principled Protection: Notice and Consent versus Collection Principles

The foregoing discussion outlines the importance of personal information as the initiator for triggering information privacy law obligations. Those obligations emanate through different privacy or fair information principles that govern the life cycle of personal information management. The principles are interlinked and do not operate on a standalone basis. The application of each principle influences the application of other principles and provides an overarching framework of protection. The basis of principled protection exists in all three of the founding jurisdictional paths of information privacy law development. However, there is a notable distinction between the United States and the other two jurisdictions about the degree of obligations applied on data collectors regarding the actual point of collection.[143] The US framework eschews broader data collection obligations and favours a more limited approach

---

[143]   van Hoboken.

entitled 'notice and consent'. Alternatively, the EU and Australia adopt specific data collection obligations that go beyond the acquisition of individual consent and require examination of the lawfulness, fairness and necessity of personal information collections.

Before we get to the differences about collection, it is first necessary to outline the principled basis of protection and how principles operate together. Bygrave has adduced eight core legal principles that reflect the fundamental aims of first-generation information privacy laws.[144] The primary principle is that personal information is to be 'processed fairly and lawfully', and this concept manifests throughout the remaining principles.[145] The lawful element is apparent – that organisational personal information collection practices must be within existing law – but the fairness criterion is more abstract in nature, particularly because general agreement about what is fair will change over the course of time. In general, the notion of fairness requires data collectors to take account of the interests and expectations of individuals who provide personal information to them. Personal data collection organisations are therefore obliged not to pressure individuals when they provide their personal information and to ensure that an individual consents to the provision.[146]

The minimality principle directs data-collecting organisations to ensure that the collection of personal information is 'limited to what is necessary to achieve the purpose(s) for which the data are gathered and further processed'.[147] Under this principle, organisations are required to collect personal information only for a relevant purpose.[148] Linked to minimality, the purpose specification principle dictates that personal information is only collected for specified, lawful or legitimate purposes and can only be used within these bounds.[149] The principle is essentially a cluster of three related subprinciples – namely, that the data collection purpose is (1) specified, (2) lawful and/or legitimate and that (3) further personal data processing is compatible with the data collection purpose.[150]

---

[144] Bygrave, *Data Protection Law* 57 referring to data protection rather than information privacy laws.
[145] Ibid 32. Birnhack 266 highlighting that the GDPR is a 'second generation' law due to its focus on micro rights.
[146] Bygrave, *Data Protection Law* 59.
[147] Ibid 32.
[148] van Hoboken 234.
[149] Nikolaus Forgó, Stefanie Hänold and Benjamin Schütze, 'The Principle of Purpose Limitation and Big Data' in Marcelo Corrales, Mark Fenwick and Nikolaus Forgó (eds), *New Technology, Big Data and the Law* (Springer 2017).
[150] Bygrave, *Data Protection Law* 61.

The information quality principle ensures that personal information is accurate, both in terms of its content and context and about the purpose of information collection and processing. The principle ensures that personal data is valid because it describes unambiguously what it pertains to and is relevant and complete with respect to the purposes of intended processing and use. Information quality requires the participation of individuals to ensure that information held is up to date. Accordingly, the individual participation and control principle is pivotal, because it ensures that persons have a measure of influence over the processing of their personal information by organisations and individuals.[151]

However, most first-generation information privacy laws do not refer to the principle directly. Instead, legislation implicitly acknowledges the principle in legal rules that govern the collection, storage and use of personal information in accordance with individual knowledge and consent.[152] Likewise, first-generation laws rarely state the disclosure limitations principle directly but implicitly require data-collecting organisations to restrict the disclosure of personal information within the confines of how data is collected and within the consent provided by individuals or by the authority of a given law.[153] The two remaining principles – information security and sensitivity – protect the integrity of personal information through the provision of adequate methods of security, particularly regarding sensitive information, which may require controls that are more stringent.

The links between the principles give rise to structures of life-cycle protection are relatively evident. For example, if a data-collecting organisation is obligated to only collect personal data for a defined purpose, then it is restricted by what data it can collect and less able to find out further details of an individual's circumstances. Similarly, if personal information can only be used for the purposes for which it is collected, then the individual providing personal information is more likely to retain control over how their information is used. Quality obligations and participation rights ensure enhanced transparency mechanisms for processing and decision making. Information security requirements minimise unauthorised disclosures of personal information. Anonymisation and deletion of unwanted or out-of-date personal information similarly assist to reduce information security risks.

At the heart of this principled interplay is adequate forms of notification to individuals about why personal information is being collected

---

[151] Ibid 63.
[152] Ibid.
[153] Ibid 67.

and how it will be used. Notification is consequently an important part of collection processes and serves as a broader transparency mechanism that permeates the whole of the personal information exchange life cycle. This permeation also alludes to the vital importance that the eponymous privacy policy plays as part of the broader exchange process. The privacy policy is at the heart of the fair and lawful processing requirement of information privacy law. Data-collecting organisations are required to notify individuals about the purpose of collection and use of personal information, in order that an informed person can make a rational decision about whether or not to provide their personal information. The making of the rational decision is based significantly on the contents of organisational privacy policies, as part of what Solove terms a structure of 'privacy self-management',[154] as outlined further in Chapter 7.

The individual consent provision is at the heart of privacy self-management; hence the privacy policy's vaunted position in information privacy law.[155] However, the issue of consent, particularly at the point of personal information collection, is also one of the key defining differences between the US approach and those of other jurisdictions. The primary protective mechanism of US information privacy law is colloquially termed 'notice and consent'.[156] This mechanism, as detailed later in application, is at the heart of the sectoral manifestations of US information privacy laws[157] and is contentious, especially in comparison to the collection obligations that arise in comprehensive jurisdictions.

All forms of information privacy law have some notification element. Notice, in this regard, is directly related to individual consent. If consent is to be deemed generally viable, it must, at the very least, be informed.[158] In other words, the point of notification prior to collection is to facilitate individual understanding and to provide the information basis for a reasoned, free and thus rational decision about provision of personal information.[159] The ability of individual personal information providers to make a rational decision based on the information presented to them in a privacy policy has been a point of significant contention, as

---

[154] Solove, 'Privacy Self-Management and the Consent Dilemma'.
[155] Austin, 'Re-reading Westin' 59.
[156] Joel Reidenberg and others, 'Disagreeable Privacy Policies: Mismatches between Meaning and Users' Understanding' (2015) 30 Berkeley Technology Law Journal 39, 43–4.
[157] Austin, 'Re-reading Westin' 72 comparing the basis of notice and consent with Westin's original notion.
[158] Daniel Susser, 'Notice after Notice-and-Consent: Why Privacy Disclosures Are Valuable Even If Consent Frameworks Aren't' (2019) 9 Journal of Information Policy 37, 47.
[159] Ibid 48.

discussed in Part III. For now, though, the focus is on the regulatory locus of jurisdictional approaches and how the United States, the EU and Australia treat collection obligations in different ways.

US information privacy laws have limited obligations for data collectors in relation to collections of personal information from individuals. There is an absence of minimality and purpose specification obligations.[160] There is no legislative or regulatory restraint that attempts to restrict the amount of personal information collected to meet a specified purpose. Instead, the purposes of organisational personal information use tend to be specified directly in law or in regulatory rules. Consequently, the limited legal focus of US information privacy law, at the point of collection, tends to regard form requirements for notification. In other words, the notice element is about ensuring that appropriate forms of notice are required to meet specified legal uses of various types of PII.[161] Again, the sectoral nature of the US framework is an important driving feature behind the singular focus on notification requirements.

Take, for example, notification requirements under the GLBA. It was noted previously that comprehensive systems have dedicated information privacy regulatory authorities. The US sectoral system does not have such a regulatory body. Instead, the FTC has undertaken the de facto role of privacy regulator in relation to its oversight of various statutes.[162] The GLBA was enacted in the late 1990s as a response to broader reforms in the financial services sector and to regulate the persistent concerns involving consumer financial privacy. As part of its operation, the GLBA authorised the FTC to develop a privacy rule that details the information privacy obligations for financial institutions[163] that collect non-public personal information and those that receive it.[164] The GLBA privacy rule has three information privacy–related purposes regarding notice about privacy policies and practices, disclosures of non-public information to third parties and an opt-out disclosure option for consumers. The rule then details in some significant depth the form and specification of notification requirements.

---

[160] MacCarthy 56 discussing the inappropriateness of minimalisation in relation to Big Data.

[161] Solove, 'Privacy Self-Management and the Consent Dilemma'.

[162] van Hoboken 245; Daniel J Solove and Woodrow Hartzog, 'The FTC and the New Common Law of Privacy' (2014) 114 Columbia Law Review 583; Woodrow Hartzog and Daniel J Solove, 'The Scope and Potential of FTC Data Protection' (2015) 83 George Washington Law Review 2230.

[163] A financial institution is one that is 'significantly engaged' in financial activities.

[164] The FTC also has mandate over the GLBA security rule, which separately sets information security obligations.

A similar structure also exists under HIPAA, with default regulatory authority vested in the US Department of Health and Human Services for the passing of and compliance with the law's privacy rule.[165] The rule applies to covered entities – namely, health plans, healthcare clearinghouses, and any healthcare provider that transmits protected health information[166] in electronic form in connection with transactions for which the secretary of HHS has adopted standards under HIPAA.[167] Notice requirements exist for entities covered by the rule, to provide an individual with notice of privacy practices that must contain certain forms of designated information.[168] Individuals provide opt-in authorisation as confirmation of consent, and subsequent uses or disclosures of protected health information have to specifically match a previous authorisation. Nonetheless, several exemptions exist, particularly in relation to governmental uses that can justify a disclosure without either individual consent or authorisation. Unusually, HIPAA also has a form of minimality principle in its Minimum Necessary Requirement. However, the requirement refers to confidentiality codes arising from disclosure of protected health information as opposed to personal information collection from individuals.

An enhanced notification requirement is therefore a common feature of US information privacy laws that forms the basis of notice and consent.[169] The position in comprehensive frameworks is somewhat different, especially regarding the imposition of further obligations at the point of personal information collection. These further obligations involve notions of lawfulness, fairness and necessity or legitimacy of collection requirements. However, like the situation with personal data and personal information, there are differences in regulatory approach in the comprehensive structures of the EU and Australia. A brief overview of overall regulatory ambit is supplied before a consideration of the collection principles is undertaken.

---

[165] Stacey A Tovino, 'The HIPAA Privacy Rule and the EU GDPR: Illustrative Comparisons' (2017) 47 Seton Hall Law Review 993.

[166] This is individually identifiable health information, which includes the individual's past, present or future physical or mental health or condition; the provision of health care to the individual; or the past, present, or future payment for the provision of health care to the individual that identifies the individual or for which there is a reasonable basis to believe that it can be used to identify the individual.

[167] 'Summary of the HIPAA Privacy Rule' (U.S. Department of Health & Human Services) <www.hhs.gov/hipaa/for-professionals/privacy/laws-regulations/index.html> accessed 20 June 2019.

[168] 45 C.F.R. §§ 164.520(a) and (b).

[169] Solove, 'Privacy Self-Management and the Consent Dilemma' 1882.

The GDPR explicitly regulates the processing of personal data where processing is conducted by data controllers and data processers and utilises personal data from individuals, the latter known as 'data subjects' in the GDPR. Article 4(2) construes processing broadly to encompass a wide range of automated and non-automated data processing operations including collection.[170] Controllers are the primary decision makers regarding data processing and exercise overall control over the purposes and means of processing.[171] Joint controllers reflect the situation where more than one controller controls processing. Processors act on behalf of, and only on the instructions of, the relevant controller.[172]

The Privacy Act implicitly regulates personal information processing and does not have a model of regulated process parties like the GDPR. Instead, the focus of regulatory scope involves APP entities that can be federal government 'agencies' or certain private sector 'organisations'. Information privacy obligations emanate from collecting, holding or using/disclosing personal information rather than from the processing of personal data, even though collection effectively begins a processing cycle of principled protection for most entities. As we will see, there are also differences in how the process of collection is regulated.

Under the GDPR, personal data collections are required to have a 'lawful' basis. Furthermore, under Article 5(1)(a), personal data shall be processed in a lawful, fair and transparent manner in relation to the data subject. As processing includes collection, that also means that collections of personal data must meet these requirements in relation to the data subject. On top of that, Article 5(1)(b) requires that personal data shall generally be collected for specified, explicit and legitimate purposes and not further processed in a manner that is incompatible with those purposes. Finally, Article 5(1)(c) requires that personal data shall be adequate, relevant and limited to what is necessary in relation to the purposes for which they are processed. These clauses, of course, represent the fairness, participation, purpose limitation and data minimisation principles highlighted by Bygrave as a key feature of information privacy law protections. These principles are also represented in the Privacy Act but in different ways, as outlined further on.

Like the United States, the notion of consent is central to the application of the GDPR. However, consent is considered within a

---

[170] Processing is 'recording, organisation, structuring, storage, adaptation or alteration, retrieval, consultation, use, disclosure by transmission, dissemination or otherwise making available, alignment or combination, restriction, erasure or destruction'.

[171] Information Commissioner's Office (UK), *Guide to the General Data Protection Regulation (GDPR)* (2018).

[172] Ibid.

broader framework of rights rather than as part of a series of specific notification requirements found in individual information privacy statutes. For example, Article 8(2) of the EU Charter specifies that personal data may only be processed based on individual consent or under another legitimate basis in law. Consent is therefore one of six foundational legal bases for the processing of personal data. The others are necessary fulfilment of contractual performance, compliance of controller-specific legal obligations, protection of the vital interests of a data subject or other person, exercise in the public interest or on behalf of official authority and whether the controller can rely on legitimate interests that do not override the interests or rights of a data subject. Consent is, not surprisingly, construed broadly and explicitly as

any freely given, specific, informed and unambiguous indication of the data subject's wishes by which he or she, by a statement or by a clear affirmative action, signifies agreement to the processing of personal data.[173]

In keeping with the notion of fairness and transparency that runs throughout the application of data processing in the GDPR, consent must be explicit and cannot be bundled together with other interests to obfuscate individual understanding and thus reduce its meaningfulness.[174] The importance of consent as a foundational ground for data processing is acknowledged by the fact that controllers need to retain evidence of consent declarations that can be matched to purposes of use and that a data subject can withdraw consent at any time through an easy means provided by the controller. However, it is currently unclear as to how the designation of 'freely given' will be construed in the context of online services that equate consent with a 'take it or leave it' choice for the data subject.[175]

Consent is also a factor in the application of the Privacy Act but is considered in a less foundational manner, especially in relation to collections of personal information. The relevant APPs cover the same foundational principles – namely, fairness, participation, purpose limitation and data minimisation – but they do so outside the central scope of processing. Instead, separate privacy principles govern different parts of the personal information exchange process, but they interlink to provide a structure of comprehensive coverage that attempts to ensure fair application throughout. The collection process is covered by three principles: APPs 1, 3 and 5.

---

[173] Article 4(11), GDPR.
[174] van der Sloot and Zuiderveen Borgesius.
[175] Ibid.

APP 1 is essentially a public declaration of how an entity will handle personal information supplied to it. The more an APP entity is open and transparent about its personal information handling practices, the higher the level of certainty an individual will have regarding the handling of their personal information. In that sense, APP 1 is the underlying basis for the imposition of fairness in the personal information exchange process. APP 1 has six elements relating to open and transparent behaviour. An APP entity is required to take reasonable steps to implement practices, procedures and systems to ensure that the entity complies with the APPs, and to deal with related inquiries/complaints (APP 1.2). It must also have a clearly expressed and up-to-date privacy policy (APP 1.3 and 1.4) and take reasonable steps to provide a privacy policy free of charge (APP 1.5) and in the format requested by an individual (APP 1.6). APP 1.2 is particularly important, as it encourages entities to be proactive about their levels of privacy obligation by requiring APP entities to state clearly what their privacy-related obligations are and how it will meet them. APP 1.2 is intended to focus the corporate mind and create an acknowledged standard upon which the organisation will be held accountable. If the organisation breaches APP obligations, then a complaint can arise.

The brief coverage of APP 1 highlights that the construct of reasonableness is much greater in the Privacy Act compared to the GDPR and the US information privacy laws. 'Reasonable' is not defined in the Act, and the term is generally used to qualify a test or data collector obligation. In general, 'reasonable' is used as a foundation for an exemption, where an APP entity must do something that may be qualified if it is unreasonable to expect the entity to carry out the action. Reasonableness is consequently a key method in balancing competing fairness interests between individuals and APP entities. 'Reasonable' has the following components: it is based on ordinary meaning, according to reason, and capable of sound application; it is also based on fact, and therefore the application of reasonable steps is inherently contextual, and so 'reasonableness' is defined within the context of case-by-case, factual application and an objective test – namely, how a reasonable person who is properly informed would be expected to act in the circumstances.[176] This test of reasonableness is influenced by current standards and practices. The OAIC guidelines are important in this regard, as they set parameters for what will be considered reasonable in the context of personal information exchange.[177] It is then up to the entity to show that its actions were reasonable.

[176] Office of the Australian Information Commissioner, *Australian Privacy Principles Guidelines* (OAIC 2014) 23.
[177] Ibid 8.

APP 3 specifically covers the collection of solicited personal information, which means personal information collected directly from an individual. Collection is defined relatively and takes place when an entity gathers, acquires or obtains personal information from any source and by any means. APP 3 essentially involves two elements: when an entity can collect personal information and how an organisation can collect personal information. There are different obligations for public sector agencies and private sector organisations, and this book focuses on the latter. There are three obligations for organisations that flow from APP 3: necessity requires an organisation to only collect personal information that is reasonably necessary for one or more of its functions or activities (APP 3.2); collections of sensitive information require consent or have to fall under an exemption (APP 3.3); the collection is lawful and fair (APP 3.5) and directly from the individual unless an exception applies (APP 3.6).

Whether a collection is reasonably necessary to fulfil a function or activity requires examination through a two-step process. First, it is required to identify the organisation's functions and activities. These will be what the organisation currently does or what it plans to do in the future – in other words, what the organisation does as part of its daily activities as expressed in public documents, such as its website or annual report. Second, a determination is required to assess whether a collection is reasonably necessary for one of those functions or activities. This is where the objective test of reasonableness becomes important; namely, a reasonable person who is properly informed would agree that the collection is necessary. It is up to the organisation to justify that the collection was reasonably necessary.

APP 3.5 requires that personal information collections be lawful and fair. The first point to note is that both 'lawful' and 'fair' are undefined in the Act. That said, a *lawful collection* is generally easier to identify. Put simply, a lawful collection is one that is not prohibited by law. Law in this sense can be a statute, a regulatory rule, a civil wrong or a court order. A fair collection is a bit more abstract and is not as straightforward. A fair collection does not involve intimidation or deception or involve collection that is unreasonably intrusive.[178] What is an unfair or a fair collection is inherently contextual. It depends on the circumstances of each individual collection. Generally, a covert collection will be one that involves deception, so it is more likely to be unfair. However, some covert collections in certain contexts may

---

[178] MacCarthy 61 defining fairness in a different context of accuracy, discrimination and transparency in an operational sense.

be required, such as in benefit or fraud investigations where to notify the individual about the collection would defeat its purpose. In these circumstances, it could be reasonable for a seemingly unfair collection to be fair.

Linked to both the application of APP 1 and APP 3 is the process of notification under APP 5. As noted previously, notification plays a fundamental role in information privacy, and it is at the heart of privacy self-management. This point is represented in APP 5.1, which states that an 'entity must take reasonable steps either to notify an individual of the "APP 5 matters"[179] or to ensure that the individual is aware of those matters'. There is a distinction here between 'notify' and 'aware' that orients to a different form of consent to the GDPR. 'Notify' means that an individual was directly notified through a specific notice (a form) at the point of collection, whereas 'aware' relates to the situation where an organisation only provides access to notification. The latter can take the form, for example, of a hyperlink to a notice on a website. It is important to note that both are acceptable forms of notification.

However, notification as a 'reasonable step' requires a contextual understanding of the collection. The types of considerations involved in reasonable steps include the type of information collected, the adverse consequences that could flow because of the collection and the special needs of the individual. Organisational inconvenience and cost, on its own, will not be a factor that amounts to a reasonable step. As such, an organisation cannot simply claim that notice would be too expensive and therefore it cannot be given. The barriers to notice would have to be excessive in all the circumstances, in that it would be structurally difficult for an organisation to give notice.

As noted earlier, notification is all about the promotion and facilitation of informed choice regarding the provision of personal information. Consequently, for an informed choice to occur, a data collector should notify an individual at or before the time it collects personal information, unless it is not practicable to do so. Otherwise, the notification must take place as soon as practicable after the collection. The test of practicability is an objective test, and it is up to the organisation to justify that notification was not practicable and that a delay was necessary. That said, the general expectation, in keeping with privacy self-management, is that individuals should be adequately notified of collections in advance or at the time of collection. However,

---

[179] APP 5.2 details several information requirements that need to be provided.

no notification or no attempt to make an individual aware can equate to a reasonable step in certain circumstances. As such, the Act recognises that some collections do not generate notification expectations – for example, where an individual should be aware of the purpose of collection, such as where the collection is regularly recurring and there is no need to notify on each occurrence.

Clearly, notification plays an important role in garnering consent. Yet, despite the detailed requirements of APP 5, unlike the GDPR, which places express forms of consent as a foundational legal basis for processing, express forms of consent are not required for collections of personal information under the Privacy Act.[180] Implied forms, where consent can reasonably be inferred from the conduct of the individual and the APP entity, are acceptable. Even though the Act construes consent broadly,[181] it is heavily linked to the justification or establishment of an organisation's exception to a principle. Consent is an exception to a general prohibition against personal information must be handled in a certain way. An entity can do something with collected personal information that would normally not be allowed, if consent has been given by an individual, including through inferred and implied forms.

Both the GDPR and the Privacy Act have wide-ranging exceptions that also operate in different ways. The GDPR excludes four situations of processing activity, for purely personal or household activities, for national security purposes that are necessary and proportionate, for freedom of expression and for archiving purposes in the public interest, particularly journalism.[182] The Privacy Act's exemptions, on the other hand, focus on types of entity and personal information records.[183] Federal law enforcement, courts and national security agencies are exempt, as are political parties, including volunteers and contractors undertaking political acts for electoral purposes. Equally controversial, in the private sector, small and medium-size enterprises with an annual turnover of less than $3 million (Australian) are exempt, as are media organisations for acts undertaken in the course of journalism. Most private sector employee records that hold personal information involving a current employment relationship are also exempt.

---

[180] They are however required for collections of sensitive information.
[181] The OAIC considers consent as four components: (1) informed; (2) voluntary; (3) current and specific and (4) capacity.
[182] van der Sloot and Zuiderveen Borgesius.
[183] Moira Paterson and Maeve McDonagh, 'Data Protection in an Era of Big Data: The Challenges Posed by Big Personal Data' (2018) 44 Monash University Law Review 1.

Like the GDPR, the Act does not apply to individuals in a non-employment capacity. Section 16 states that nothing in the APPs applies to the collection and use of personal information by an individual, or personal information held by an individual, for the purposes, or in connection with personal, family or household affairs. The Act therefore only applies to individuals who are conducting acts or practices in relation to their employment rather than personal life. Moreover, the Act does not apply to an organisation that is an individual – for example, a sole trader – if the act or practice was not conducted in the course of business. Taking on board the coverage of exceptions, the comprehensive coverage of the GDPR is much greater than that of the Privacy Act – a point that is discussed further in the final part of this book.

One final point about the difference between jurisdictional approaches needs consideration. Both the GDPR and the Privacy Act have extra-territorial scope. Under Article 3, the GDPR can apply to non-EU companies that are outside of the EU when (a) those companies offer goods and services in the EU, whether free or paid (for example, a website or online service, such as cloud storage, that can be used by EU citizens) or (b) those companies monitor the behaviour of individuals in the EU (for example, a social network that has EU members or allows EU members to join). The effect of Article 3 has yet to be fully appreciated. However, Chapter 4 signified its effect through the changes of smart home device privacy policies, including those of US manufacturers Canary and Nest, following the implementation of the GDPR.

Section 5B(1A) of the Privacy Act has a more constrained notion of extra-territorial operation. An organisation that conducts an act or engages in a practice outside of Australia is covered by the Act if the organisation has an Australian link. An 'Australian link' considers several factors that relate to Australian citizenship, legal recognition in Australia and management and control operations. The clause is primarily designed to give the Privacy Commissioner enhanced options of investigation about individual complaints relating to personal information–handling practices conducted overseas. As such, it is not designed to have the same structural coverage as Article 3. This is evident by the fact that an overseas act or practice will not be a breach of an APP if it is required by an applicable foreign law. There are two different policy positions at play. Article 3 extends the application of the GDPR across the globe, and Section 5B(1A) expands the reach of complaint and investigation mechanisms. Note also that none of the US information privacy laws detailed in this chapter have extra-territorial effect, which again reflects the limited ambit of the notice and consent model.

6.5   Conclusion

Part II outlines the legal framework for considering the collected challenges that will arise. Chapter 7 will examine those challenges in greater depth in relation to the smart home and the insurance business models highlighted in the Part I. Before we get there, some points need to be noted about the conceptual and practical coverage. Chapter 5 highlighted the intersecting conceptual foundations of information privacy. Even though, as evident from this chapter, information privacy law is overtly founded on control concepts, the other conceptual implications of informational space for autonomous growth, relational and social foci and power relations are still equally applicable to its effective operation. These points are explored in much greater depth in Chapter 8.

This chapter has also demonstrated the different jurisdictional loci of information privacy law. Even though US, EU and Australian information privacy law can all be traced to similar principled roots, their ultimate manifestations have taken essentially different tracks, as highlighted in the coverage of regulated information and the application of collection principles. Three underpinning policy themes emerge. The United States tends to favour a market-based approach to information privacy that has the effect of a narrower construction of PII to meet a specific statutory purpose. Collection protections are equally constrained with regard to notification requirements for consent acquisition. The EU favours a rights-based approach to information privacy – hence broad constructions of personal data and principled protections of lawfulness, fairness and transparency embedded through data processing activities, including collection. Australia, as a non-European OECD country, utilises information privacy law to provide a minimum standard of individual protection, balanced against the needs of organisational exchange. Balance is derived through core constructs such as reasonableness, which can result in lower levels of coverage and scope when compared to the GDPR. Nevertheless, a suite of collection principles exists that seeks to imbue fairness and lawfulness into the collection process.

Let's take stock before we dive deeper. We have a world that is increasingly collecting sensor data from individuals, buildings and environments. We have a fragmented data collection framework in the smart home based on sensor collections from many providers. We have rapidly evolving business models that seek to commercialise smart home sensor data in several ways. We have different concepts of information privacy that intersect, entangle and compete. We have different jurisdictional approaches to information privacy law that produce different definitions of regulated information and different levels of protection in relation to collections of personal information.

The true complexity of the collected world thus starts to unfold. That complexity is now considered further in relation to the information privacy challenges that will arise from the collected world and the conceptual and legal responses that are required to deal with these levels of complication.

Part III

Information Privacy Law for a Collected Future

7   Collected Challenges

7.1   Introduction

The final part of the book examines the conceptual and legal application of information privacy in the face of a collected world. Coverage
is given to the smart home sensor-based architectures and emerging
business models introduced in Chapters 3 and 4. Several key points
are covered. First, this chapter highlights how ubiquitous collections
of sensorised data will challenge the conceptual basis of information
privacy and its legal manifestation as a process of principled protections for specific categories of regulated information. Next, Chapter 8
draws principally upon Julie Cohen's work, particularly her concept of
modulated power, to further critique information privacy law's control
basis. Chapter 9 then puts forward some design points and suggestions
for information privacy law reform, in keeping with a reformulated perspective of information privacy law based on Cohen's work. Finally,
Chapter 10 concludes the book and asks whether we are heading
towards a smart, a collected or a modulated world.

Before we get that far, it is first necessary to articulate the challenges arising from the collected world. Collected challenges are outlined from several perspectives. First, building on the smart home
coverage in Chapters 3 and 4, this chapter identifies the information
privacy concerns that arise from academic considerations of the smart
home. Several considerations are worth note. The smart home literature
tends to consider issues of privacy in a smart home context predominantly in relation to information privacy concerns regarding collections
of data about smart home inhabitants. Privacy risks, in this sense, are
constructed as a combination of information privacy and information
security issues. Another common feature regards compliance with established information privacy principles. However, there has generally been
little attempt to address the complex legal question of what data collected from a smart home would be classifiable as personal information.

Most research has simply referred to the information privacy issues arising from all forms of data from the smart home.

Furthermore, the predominant focus on information privacy as a concern of control is problematic because of the collected challenges that arise from upon sensorised collections, as evident from the smart home coverage in Part I. Sensorised collections challenge the very basis of information privacy law protections. The type of data collected from sensorised devices starts to push the bounds of classificatory strategies predicated on being able to identify an individual from singular or aggregated forms of data. Sensorised data is principally data about the functioning of a device that can record environmentally focused activities within the specified range of the sensor. As highlighted in the anatomy of the smart home, different sensors have different ranges and can therefore collect different varieties of environmentally focused data. The types of sensorised collections in the smart home thus give rise to significant classificatory challenges. The classification process of personal information becomes more challenging because different strategies of classification can result in alternative constructions of personal information. As such, different jurisdictions are likely to come to different results regarding the issue of whether, when and how smart home event data is to be regulated by information privacy law.

It also must be remembered that the type of sensorised challenges that emerge from collected environments such as the smart home are done in highly fractured and contested data collection environments. Another challenge therefore regards the transition from binary to multiple data collection environments, which stretches the functional bounds of information privacy law processes.[1] The infrastructural architecture of the smart home means that data collection and exchange is conducted by multiple parties rather than a singular entity. Chapter 3 highlighted the complexity of data collection environments in which there are multiple collectors, multiple routes of collection and multiple purposes for collection. Binary processes are therefore challenged by the multiple nature of these fractured environments.

Collected challenges also arise from the mode of sensorised collection. This type of collection does not have the same purposeful

---

[1] Mark Burdon, 'Privacy Invasive Geo-Mashups: Privacy 2.0 and the Limits of First Generation Information Privacy Laws' [2010] University of Illinois Journal of Law, Technology and Policy 1.

intention that is pertinent to the traditional focus of information privacy law. Instead, it relies on the passivity, rather than the agency, of individuals as an intrinsic part of the purposeful process. Sensorised collections thus challenge the principled basis of information privacy law's protective structures. This can be seen through the application of key principles, particularly the collection principle, as demonstrated by the privacy policies of the smart home insurance models covered in Chapter 4. The policies exhibit different collection strategies which influence the notice and subsequent disclosure processes adopted. All of this leads to questioning the control basis of information privacy, particularly in environments that are increasingly boundary reduced.

7.2   The Limits of Identified Privacy Concerns

Privacy concerns discussed in the smart home literature have tended to be conceptualised as pure information privacy or data protection issues. For example, a study involving smart home academic researchers identified privacy as a key issue, but it was often conceived as 'the unbidden sharing of data, rather than as an invasion of personal space'.[2] Information privacy issues have thus focused on collections of data from smart home inhabitants and tended to relate to specific risks, particularly construed as data breaches.[3] Information privacy risks, in this sense, have been constructed also as information security issues that could arise from a series of similar situational contexts, such as the use of certain data storage practices coupled with unknown forms of data accessibility, particularly in a commercial context.

Another common feature regarded compliance with established information privacy principles. However, few papers have attempted to address, in any significant depth, the complex legal question of what data collected from a smart home would be classifiable as personal information. This classification is the threshold issue for the application of information privacy, as noted in Chapter 6. Any smart home data that is not classed as personal information will not be covered by information privacy law. Most papers have simply referred to the information privacy issues arising from all forms of data coming from

[2] Giles Birchley and others, 'Smart Homes, Private Homes? An Empirical Study of Technology Researchers' Perceptions of Ethical Issues in Developing Smart-Home Health Technologies' (2017) 18 BMC Medical Ethics 1, 4.
[3] Geneiatakis and others; Marianthi Theoharidou, Nikolaos Tsalis and Dimitris Gritzalis, 'Smart Home Solutions: Privacy Issues' in Joost van Hoof, George Demiris and Eveline JM Wouters (eds), *Handbook of Smart Homes, Health Care and Well-Being* (Springer 2014).

the smart home. Such a perspective may be a sensible policy outlook to embark from, but it does not assist in identifying the complexities of smart home privacy considerations, even in the more straightforward realm of information privacy as a control mechanism.

Privacy concerns are therefore largely considered within the traditional focus of information privacy law as issues of data collection, storage and use. Moreover, and perhaps not surprisingly, information privacy solutions for the smart home are predicated upon enhanced forms of individual control over smart home data infrastructures and components. These forms of control have been largely technical in focus – for example, the development of a 'Privacy Controller' that considered data collection in conjunction with the privacy preferences of individual users.[4] Another common form of control protection features stronger forms of encryption, particularly at the device level.[5] Other control mechanisms have included a gateway architecture[6] and the development of specific privacy risk assessment models to highlight areas of individual losses of control.[7] The complexities of implementing individual privacy solutions across device, communication and platform levels has been seen as a significant problem, particularly as a single manufacturer could not guarantee security across the complete smart home.[8] Corporate information privacy and security responsibilities, entailed in data collections from the smart home, are consequently unclear and are likely to remain complex in a fluid and developing environment. These types of environment typify the future spaces of the collected world.

The issue of user consent has also been a persistent sub-theme in the discussion of improved forms of user control in smart homes. A key challenge that arises from the smart home regards how to appropriately inform inhabitants about the complex forms of sensorised data collection, and in particular decision outcomes based on predictive analytics.[9] The issue of appropriate consent mechanisms is therefore a difficult and challenging issue to resolve, but it is one that smart home data collectors should pay particular attention to given the degree of reputational risk arising from a perceived misuse of customer data.[10] Moreover, given the number of data-collecting parties involved in smart home data collections, making customer data available to other

---

4 Latif and others 65.
5 Bugeja, Jacobsson and Davidsson; Risteska Stojkoska and Trivodaliev.
6 Lin and Bergmann.
7 Jacobsson and Davidsson.
8 Barnard-Wills, Marinos and Portesi.
9 Jacobsson, Boldt and Carlsson, 721.
10 Barnard-Wills, Marinos and Portesi 45.

parties in order to provide a seamless smart home experience could be 'fraught with difficulty',[11] particularly regarding the transition from primary to secondary uses.

As noted already, such notions of 'privacy difficulty' are largely considered through the perspective of a traditional information privacy lens. A key element of this perspective is the relationship between information privacy and information security. A prominent theme is the consideration of information privacy and information security concerns synonymously. A certain technical and rational logic pervades. If a smart home has better information security, then the corollary of better information is that the smart home will provide better information privacy. In this sense, information privacy and information security are rightly treated as being interlinked. However, although the two issues are considered synonymously, information privacy has appeared to be treated as a subset of information security. The largely technical nature of most smart home papers is perhaps a key factor, especially as the legal academy have seemingly yet to turn full attention to the privacy implications of the smart home, as noted later.

The concurrent analysis of information privacy and information security has resulted in considerations of the complexities of securing the smart home. The concentrated convergence of a number of different, interconnected technologies means that the smart home will be a point of intense data collection that will facilitate ever-finer profiling of individuals.[12] However, at the same time, the devices that make up the smart home are not likely to have the same degree of security as non-sensorised products, given their scale, cost and size.[13] Moreover, the number of applicable communications protocols present in the smart home (e.g. Wi-Fi, Bluetooth, ZigBee, etc.) increases the number of opportunities for security exploitation. The complex interaction of different product manufacturers, content service providers and distribution channels gives rise to some significant questions about who bears the responsibility for securing the smart home across a range of parties including device manufacturers, platform connectors and individual home inhabitants.[14]

A number of threats could consequently emerge from a vast number of smart home assets, including physical threats, equipment failures

---

[11] Balta-Ozkan and others, 'Social Barriers to the Adoption of Smart Homes' 369.
[12] Scott J Shackelford and others, 'When Toasters Attack: A Polycentric Approach to Enhancing the Security of Things' (2017) 2017 University of Illinois Law Review 415.
[13] Scott R Peppet, 'Regulating the Internet of Things: First Steps Toward Managing Discrimination, Privacy, Security, and Consent' (2014) 93 Texas Law Review 85.
[14] Barnard-Wills, Marinos and Portesi 40.

or malfunctions and privacy threats in the form of unauthorised uses of personal information and surveillance consequences.[15] A particular threat regards the vulnerability of smart home devices to internal and external hacking attacks.[16] For example, in one study, researchers tested the security capabilities of twenty popular smart home devices and concluded that each device exhibited some form of vulnerability.[17] Some devices were palpably more secure than others, and some products were susceptible to attack in a number of different ways. As noted already, risk analysis frameworks for information security and information privacy in the smart home have begun to be developed, particularly in relation to the poor security characteristics evident in smart home devices. The development of risk assessment frameworks is complicated by the fact that different smart home residents may have different risk appetites regarding information security, information privacy and convenience.[18]

Although there is a subtext within the general smart home coverage that broader privacy concerns do exist, the concerns are not quite fully articulated, particularly given the technically heavy orientation of the literature. In these considerations, information privacy is viewed as something that relates to an expansive concern which is difficult to quantify. For example, a general concern arises regarding restrictions of individual autonomy and freedom, similar to the conceptual considerations of information privacy as a protector of personal autonomy, as noted in Chapter 5.[19] These autonomy concerns are deemed particularly pertinent in the type of fully automated and semi-autonomous decision-making capacities of Aldrich's attentive home, as detailed in Chapter 3, in which smart home infrastructures automatically adjust device preferences with little or no input from individuals. The ability of individuals to undertake freely made decisions is thus questioned, and the concern arises that smart home inhabitants will be controlled by nudging apparatuses without realising it.[20]

---

[15] Ibid 43.
[16] Noah Apthorpe and others, 'Spying on the Smart Home: Privacy Attacks and Defenses on Encrypted IoT Traffic' https://arxiv.org/pdf/1708.05044.pdf; Nikole Davenport, 'Smart Washers May Clean Your Clothes, but Hacks Can Clean Out Your Privacy, and Underdeveloped Regulations Could Leave You Hanging on a Line' (2016) 32 The John Marshall Journal of Information Technology & Privacy Law.
[17] Sivaraman, Gharakheili and Fernandes.
[18] Wilson, Hargreaves and Hauxwell-Baldwin, 'Benefits and Risks of Smart Home Technologies'.
[19] D Townsend, F Knoefel and R Goubran, 'Privacy versus Autonomy: A Tradeoff Model for Smart Home Monitoring Technologies' (Annual International Conference of the IEEE Engineering in Medicine and Biology Society, Boston, Mass, 30 August).
[20] Richard Harper, *Inside the Smart Home* (Springer 2003).

Similarly, there is tacit consideration about potential surveillance consequences that could arise from being monitored and tracked in our most private spaces. The purveyors of surveillance are themselves numerous and can cover the full gamut of corporate data collectors and different government departments, especially law enforcement and national security agencies.[21] Also a concern is the use of sensorised devices for surveillance purposes by smart home inhabitants, in which the smart home is used by one inhabitant to monitor and track the behavioural activities of another.[22] The multiplicity of surveillance operators and opportunities could therefore mean that the smart home will be a prime site of development as part of a 'big brother' form of existence.[23]

This brief coverage of the information privacy issues that arise from the smart home is important to this chapter's overall consideration. For the most part, information privacy issues are focused upon traditional forms of information privacy law or are related to other privacy law elements. However, a couple of points are of clear importance. Sensorised data collections are vital to the broader operation of the smart home and the developing smart components of the collected world. However, while sensorised data collections are key to the structures of unfolding smartness, the actual processes of data collection are challenging due to the fragmented nature of infrastructural frameworks of the smart home and the collected world.

7.3   Sensor Collections are Different

Sensor data collections have already played a significant part in this book's coverage. Chapter 2 highlighted that sensor data collection is one of the intrinsic features of the collected world. Chapter 3's anatomy of the smart home highlighted the sheer density of sensorisation required for the smart home to operate effectively. Sensor data collections are significant in the context of the wider operation of the predictive and prescriptive analytic frameworks that form automated responses in the smart home. Chapter 4 then discussed the actualisation of commercial use of smart home sensor data and highlighted how different smart home insurance models are evolving based on emerging exchange partnerships of sensor data.

---

[21] Ferguson 819–21.
[22] Zeng, Mare and Roesner.
[23] Balta-Ozkan and others, 'Social Barriers to the Adoption of Smart Homes' 369.

The advent of sensorised collections as the predominant basis of future business and partnership models will cause major challenges for the application of information privacy law. Chapter 6 detailed some key considerations about the law's operation. For example, information privacy law will only attract regulatory consequences relevant to classifiable types of sensor data that can be categorised as personal information. The simple but threshold question of what sensor data would be personal information is itself complex. Putting that aside for now, however, there are also some fundamental differences between the collection and use of sensor data compared to the types of data that would give rise to legal obligations under the traditional focus of information privacy law.

Underpinning sensorised data collection are two inherent and fundamental tensions in the collected world. Sensorised data collections are necessary to create an historical understanding of how the smart home is being utilised. The historical reconstruction of smart home usage is then reconfigured for future home automation interventions that attempt to make the home more efficient and effective.[24] However, the nature of sensor data collection means that the sensorised instrument has to become so embedded that it captures individuals in their most natural state, in their most private setting.[25] Otherwise, if individuals know that a sensorised device is present, then they will respond differently, and the data captured will be an inaccurate representation of the behaviour.[26] In that case, the smart home will not be able to respond effectively to inhabitant wishes or accurately reflect smart home usages. Accordingly, the seemingly simple act of locating sensorised data-collecting devices in a smart home could have considerable importance in relation to the identification of individual patterns of behaviour. This point is complicated further by the multiplicity of devices in the smart home through which multiple streams of data collection, aggregated together, create the possibility of revealing sensitive patterns of behaviour[27] that can also be used to identify sensitive

---

[24] Lewis 282.
[25] Peter Tolmie and others, 'Towards the Unremarkable Computer: Making Technology at Home in Domestic Routines' in Richard Harper (ed), *Inside the Smart Home* (Springer 2003) 200.
[26] John Strain, 'Households as Morally Ordered Communities: Explorations in the Dynamics of Domestic Life' in Richard Harper (ed), *Inside the Smart Home* (Springer 2003).
[27] Barnard-Wills, Marinos and Portesi 8.

aspects of an individual's life or personality,[28] including information that would normally not be disclosed to other persons.[29]

All of this indicates that sensorised collections are different to the types of data collection activities upon which information privacy is historically based.[30] Sensorised collections, then, shift from decision-making activities that involve individual agency to forms of passive supplication.[31] Traditional processes of personal information collection are based on the conscious acts of autonomous individuals. The privacy policy or collection notice is used as a beacon of awareness to inform a data-providing individual about the positive and negative consequences of data collection. The beacon is intended to heighten awareness about the consequences of data collection activities so that the subjects can reassert individual control and thus preserve informational space in decision making that protects autonomous growth. Whether or not one agrees with the ability of a privacy policy in achieving these lofty aims, the awareness-raising mechanism nevertheless is a foundational element of privacy self-management that pervades information privacy law.

However, sensor collections require the opposite. Regardless of the effectiveness (or not) of a privacy policy in raising awareness, the profiling inferences based on sensor data collections require individuals to be a passive target. As already noted, passive collections are an intrinsic feature of the accumulation of accurate sensor data for individual or environmental profiling purposes. In other words, sensors can capture elements of our innermost patterns that we ourselves are not able to perceive. In turn, these perceptive insights require us to be at our most natural, which means that we need to be unaware of sensor data collections.

A perverse effect regarding autonomous actions thus becomes the justification for passive collections. The reduction, or even elimination, of individual awareness for sensor collection is justified as forms of insight-based autonomy produced in machine-dominated infrastructures. Individuals do not really know what they ultimately want because they do not have the means to accurately monitor, track and understand their own behaviours. However, machine-based frameworks can, so corporate entities can truly determine what individuals need by knowing their activities better.[32]

---

[28] Ferguson 863.
[29] Peter Tolmie and Andy Crabtree, 'The Practical Politics of Sharing Personal Data' (2018) 22 Personal and Ubiquitous Computing 293, 304.
[30] Nissenbaum, 'Must Privacy Give Way to Use Regulation?' 13 and the discussion on 'sensory modalities'.
[31] Andrejevic and Burdon 20.
[32] Hartzog 426 regarding control and autonomy with control as an illusion.

Sensor data collections are intentionally adopted because the 'digital traces'[33] that individuals leave behind through their activities in the home, and elsewhere, can be used to build a sophisticated profile of user activities, behaviours and feelings.[34] However, there is a vast awareness gap between the dropping of digital traces regarding the environmental use of certain data-collecting objects and how those traces can then be used to formulate profiles of individual and environmental behaviours and activities.[35] There is an even bigger gap when such data is to be used to profile and predict moods, emotions and feelings. The awareness gap, as highlighted in Chapter 8, is simply too big for the process of notice and consent to ameliorate.[36] It is not an issue of better forms of information provision as part of a move to create a rationally enhanced data transaction process. Instead, there needs to be a much greater recognition that sensor collections are different from traditional collections of personal information and ultimately challenge the control and autonomy basis of information privacy.

Similarly, another key difference with sensorised data is the move from specific moment-in-time collection to always-on collections, which again challenges the control and autonomy bases of information privacy law. As noted previously, the control-focus of information privacy requires the stimulation of conscious human agency in data collection decision making. However, the type of decision making historically required from a data-providing individual has been relatively binary and simple, as it involved a singular informational decision. In other words, if I provide my data for this purpose, then I will receive this benefit, or I do it to fulfil this requirement. The consequential calculus underpinning the decision is therefore relatively straightforward. Information privacy's role in autonomy enablement is achieved through the vestiges of control in singular decision-making capability structures. The limited control processes at the heart of information privacy law are more effective in informational decision-making situations that involve limited choices, identifiable outcomes and self-quantifiable risks. Autonomous decision-making capacities are more likely to be possible in the sense of singular decisions.

However, the nature of sensorised collections is, again, fundamentally different to those envisaged by the control operations of information

---

[33] Tolmie and Crabtree 294.
[34] Barnard-Wills, Marinos and Portesi.
[35] Elaine Sedenberg, Richmond Wong and John Chuang, 'A Window into the Soul' in Bryce Clayton Newell, Tjerk Timan and Bert-Jaap Koops (eds), *Surveillance, Privacy and Public Space* (Routledge 2019).
[36] Gurses and van Hoboken 20; the discussion here is like that of the temporal limits on informed consent.

privacy law. Sensor data collections are continuous over time.[37] In essence, there is not a one-point-in-time individual decision to be made about conscious agreement to data provision. Instead, the utility of sensor data collections requires an accumulation of passively recorded activities. One expected decision of human agency is subsumed into multiple decisions of sensor data provision that are required to be obfuscated by structures of embedded invisibleness to create environments of unawareness.[38] The autonomous context of information privacy is thus problematic, because decision-making requirements and consequences for sensor-based collections are not singular, controllable and definitive. Instead, sensor data collections are multifarious, overlapping and not contingent on the awareness of agency of individual data providers. In fact, as highlighted already, the opposite is required for sensor data to be accurate.

The effect of passive collections is to move from provision of data *about* individuals to that of data *from* environments that include individuals.[39] The traditional process of personal information exchange involved controlled exchange of information about individuals. For example, exchange processes involved the provision of data that now typifies personally identifiable information in the US sense, such as name, identifiers, etc. There was a defined focus of exchange and specified formats of collection through legally structured forms. Data was provided about an individual to fulfil a specific purpose. Sensorised collections, in contrast, entail information processes that seek to accumulate data from individuals. The same forms of personal information covered in traditional exchanges are still applicable, but they only cover one element of data collection ambits. The form-based process of data exchange is now only relevant and supplied on sign-up of service. Chapter 4's coverage of smart home insurance data partnerships is an example.

The data collected as part of registration processes was singled out for special treatment in relation to all data exchange models because it is evident that this data will be regulated by the information privacy laws of the EU, the United States and Australia. It is clearly definable as personal information and, as such, information privacy law obligations will flow. However, as Chapter 4 also highlighted, the amount of registration data provided is miniscule in comparison to the volume of data

---

[37] Yeung, 'Algorithmic Regulation: A Critical Interrogation' 511.

[38] Nissenbaum, 'Must Privacy Give Way to Use Regulation?' 16 and the utter opaqueness of automated decision-making systems.

[39] President's Council of Advisors on Science and Technology, *Big Data and Privacy: A Technological Perspective* (White House Office of Science & Technology Policy 2014) 22.

collected by device sensors. At one time, the registration data would have been the most valuable data asset of data-collecting organisations. These are the marketing databases of old, as highlighted by Solove in the 2000s and detailed in Chapter 5's coverage of power.[40] Now that sensors reach into the world, including the home, registration data is no longer the most valuable or sought-after information about customers. Registration data provides static constituent elements concerning an individual, but it is the sensorised data about activities that enables dynamic insight of that individual.

Sensor data is also different because it enables new forms of identification. Individuals can now be identified in ways that no longer require collections and aggregations of collected registration data. That begs the question whether sensor data should be treated as personal information. That is a complex and vital question to answer, because any information that is not classified as personal information will not be covered by information privacy regulatory requirements. The three smart home insurance models covered in Chapter 4 provide an indication of the answer when applied to different definitions of personal information.

The definition of personal data under the GDPR based on 'relating to' and the underlying policy ambit of rights-based protection enables broad coverage of sensor data from smart home devices, particularly in an identifiable sense. As such, smart home event data is more likely to be classified as personal data under the GDPR. For example, Chapter 4 highlighted that the smart home sensor data collected by Neos as part of the Partnered Data Acquisition model was defined as personal data in its privacy policy as a corresponding response to the broader focus of the GDPR. In effect, smart home sensor data is classified as personal data by default, and thus broader coverage emanates. That said, the EU policy position is such that the answer to the question of whether smart home sensor data is personal data is much more straightforward, given the overarching rights-based perspective highlighted in Chapter 6. Note also, as a matter of policy principle, that the Article 29 Working Party indicated that smart home sensor data, in the context of IoT-related 'domotics', should generally be classified as personal data. Similarly, Chapter 6 highlighted that the Working Party has also made similar pronouncements regarding geolocation metadata from smart mobile devices and has stated conclusively that the concept of personal data is broad and is thus meant to be construed expansively.

[40] Solove, 'Privacy and Power'.

All of this would indicate that data controllers and processers that process personal data in the EU should class smart home sensor data as personal data under the GDPR, as a matter of policy, predicated on rights-based protection. As highlighted in Chapter 4, this seems to be the position that Neos has adopted in relation to its smart home insurance offering, as all device event data, including mobile phone sensor data, is classed as personal data in its privacy policy. Again, this reflects the core focus of the EU framework, which considers information privacy protections as a fundamental right of EU citizenship. In that sense, the construction of smart home sensor data as personal data reflects a broader rights-based perspective and protection.

However, the sectoral definitions of PII in the US system mean that it is unclear whether the smart home sensor data generated in the Partnered Intermediary and Platform Entity models would be classed as PII under the relevant sectoral laws – particularly the GLBA. As noted in Chapter 6, the sectoral approach of the US framework is fundamentally different to that of the GDPR. Rather than having a comprehensive information privacy law that applies to all types of PII, regulatory coverage is dependent upon whether data is covered by a specific piece of sectoral legislation. There are no US-sector specific laws in relation to smart home sensor data. On its face, smart home sensor data collections would simply not be covered. However, smart home sensor data collections for insurance offerings are likely to be covered by the GLBA, and if so, the FTC will be responsible for enforcing the Act's privacy rule.

As noted in Chapter 6, The GLBA aims to protect consumer financial privacy by placing limits on when financial institutions, including insurance companies, can disclose a consumer's NPI to third parties. Financial institutions must notify customers about the institutions' information-sharing practices and of their right to 'opt out' if they do not want their information shared with certain third parties. Anything that is not classed as NPI can then be disclosed to a third party without the consent of a consumer. This has important considerations for smart home insurance and is representative of a different regulatory focus – namely, the US market-based remedies that focus on facilitating the exchange mechanisms of PII rather than on individual rights-based protections under the GDPR. The US privacy policies of the Partnered Intermediary and Platform Entity business models, covered in Chapter 4, are a case in point.

The privacy policies of smart home device providers (e.g. Canary and Nest) and insurers (e.g. State Farm and Liberty Mutual) treat PII in a very different way from Neos under the GDPR. Unlike the EU policy position, smart home sensor data under the US sectoral system is not treated as PII by default, and thus more data can be exchanged between partners without information privacy law coverage. In effect, the narrow classification of PII provides narrower coverage of information privacy protections. The US privacy policies clearly differentiate between the type of information that would be classed as NPI (such as name, address, account numbers, etc.), which consumers consent to sharing, and smart home sensor data that is not classed as PII and can be shared without customer consent. This of course is a much lower form of protection, overly predicated on a model of notice and consent which has been a key criticism of the US model of information privacy and is covered further in Chapter 8. It does, however, mean that new forms of business models can develop more easily due to the ability to exchange consumer data with commercial third parties, such as illustrated in the exchange of smart home sensor data in the Partnered Intermediary model.

Two different consequences thus arise from the application of EU and US regulatory models of information privacy. In the EU, as a matter of policy, smart home sensor data should be treated as personal data by default. In the United States, it is unlikely, or at least uncertain, whether smart home event data would be classified as PII, and it is certainly not treated as such by policy default.

The issue is also unclear in the OECD system of Australia, which again highlights the effect of regulatory underpinnings with regard to the classification of information that is a target for information privacy regulation. One word can indeed make a difference, and the change of definition to focus on 'about' rather than 'relating to' has been substantial in terms of how Australian information privacy law treats smart home sensor data. Chapter 6 highlighted that the Australian situation is made more complex by the uncertain state of law at present, following a decision from the Full Federal Court in 2017. The *Telstra* case partially affirmed the previous AAT decision by confirming that 'about' does have substantive application and that therefore it is necessary to consider whether information is about an individual before assessing whether it identifies an individual. Personal information can have 'multiple subject matters', and thus an evaluative conclusion is required that considers information in its totality to determine whether it is about an individual or not. As such, even though one piece of information may not be about an individual, it can become so when aggregated with other information. It should be noted that one judge in the *Telstra* case concluded that mobile phone metadata would not be about an individual and would therefore not be personal information.[41]

---

[41] *The Privacy Commissioner v Telstra Corporation Limited* [2017] FCAFC 4 (19 January 2017) [2] Dowsett J.

At this point in time, the only thing that can be said with any degree of clarity in Australia is that collected smart home sensor data may be personal information. But it also may not be. The first question to ask is whether smart home sensor data, collected for insurance purposes, is about an individual. The issue of multiple subject matters becomes important at this point. If the *Telstra* case and the OAIC guidelines are applied, then smart home sensor data could be about an individual where data can be about smart home devices and where the 'evaluative conclusion' about aggregated smart home sensor data reveals or conveys something about an individual. The context of smart home sensor data for insurance purposes as a personalised behavioural tracking platform thus becomes important in considerations regarding subject-matter characterisations of 'about' and the subsequent identification capabilities. However, it must be borne in mind that the *Privacy Commissioner* case, and the preceding AAT decision, indicated that mobile phone metadata would not be treated as personal information. The only conclusive determination that can be made is that it is unclear whether smart home sensor data would be personal information and thus needs judicial testing.

The second question to ask is whether an individual is identified or reasonably identifiable from smart home sensor data. The degree of ease and certainty, in the context of moderate steps required by smart home event data collectors to identify an individual, becomes the key issue to now resolve. For example, the Privacy Commissioner's initial decision in the *Grubb* litigation focused heavily on the fact that Telstra had existing organisational capacities to undertake the aggregation of mobile phone metadata for law enforcement purposes.[42] As such, the Partnered Data Acquisition model, as exemplified by Neos, is more likely to have aggregation capabilities to identify an individual through reasonable steps, as personalised behavioural tracking is a significant part of that business model. Thus, the degree of certainty regarding identification is likely to be high, and the identification process steps are relatively moderate for Neos, as individual behavioural tracking is a core component of its business model. However, it is not clear to what extent this logic would apply to the Partnered Intermediary and Platform Entity models, which are not so heavily predicated on individual behavioural monitoring. Again, it is a question that would require judicial consideration as to when multiple subject matters can emanate in identification.

The uncertainties around the threshold issue regarding information privacy law application is representative of the different foci of

[42] *Ben Grubb and Telstra Corporation Limited* [2015] AICmr 35 (1 May 2015) 34–5.

regulatory models (covered in Chapter 6) and clearly exemplified in the classification of smart home sensor data as personal information. Under the GDPR, this data would very likely be personal data. Under different US sectoral laws, it is not likely to be classed under a variety of different types of PII. Under the Australian Privacy Act, it is uncertain whether it would be personal information. The analysis of the threshold issue is important, because it highlights the different applications of regulatory models of information privacy law. These models are not fixed, and in the context of the many smart home sensor data collectors that will be involved in smart home business models, this issue must be borne in mind. What also must be borne in mind is the effect of the smart home itself and that it is a space of rapid commercial development and fractured and contested in many ways.

7.4   Fragmented and Contested Environments

The first part of the book detailed the rapid onset of a collected world through the proliferation of sensorised devices, infrastructures and environments. The smart home, as a prime target of commercial interest, has been used in this chapter to consider some of the collected challenges that will arise for information privacy law. It is important to acknowledge that data collections of all types take place in a real social context. As the coverage of Nissenbaum's contextual integrity in Chapter 6 highlighted, social norms play a significant part in understanding the complexity of different social contexts involved in exchange processes of personal information. The technocratic nature of information privacy law can have the effect of divorcing the application of its principled processes from the reality of everyday life. Those processes, again as detailed in Chapter 6, outline the scope of control mechanisms accorded to individuals and the obligations placed upon data-collecting entities. However, the clearly defined processes of information privacy law do not necessarily reflect the complexity, fragmentation and contestation that arise from social processes of data collection. Take the smart home as an example.

As noted in Chapter 3, the smart home has been in an evolving state for over half a century. Technological developments appear recently to have increased significantly through the convergence of sensorised technologies, consumer Wi-Fi home architectures and platform-based data communication pathways. Despite these developments, and the importance of the home as a foundational building block of liberal societies, there is currently no coherent legal and regulatory response to the burgeoning smart home market. Smart home commercialisation developments are nevertheless gathering great speed. For example, as highlighted in Chapter 4, major insurance companies worldwide are investing significant research funds in partnerships with smart home system and device manufacturers. The market is rapidly expanding, but business models are yet to fully augment their coverage in response.

The developing smart home commercialisation spaces are therefore very fluid. The commercialisation of the smart home is a space in which numerous types of corporate actors are involved in the search for business opportunities that can translate sensor data collections into commercial value. However, it seems at this stage that no commercial actor has confidently determined or actualised how value can be created through smart home business models. Thus, no actor has really answered the question yet of how you generate value from smart home business models predicated on collections of smart home sensor data that identify and predict better risk assessments of customer behaviour. One prevailing characteristic of this fluid environment does persist: the dominant operational mode of smart home business models appears to be partnerships between established corporate service and product providers, such as insurers, and smart home service or device providers.

The uncertain search for value through partnership arrangements based on smart home sensor data reflects the rapid evolutionary development that has taken place, which has created fragmentation at every conceivable level in the smart home environment. The complexity of smart home technological infrastructures and networks compound fragmentation. The only apparent constant in this environment is the drive for the cumulative generation and acquisition of smart home sensor data. As noted throughout the book, this data is key to business model smart home business model development and is the source for future partnership intentions based on structures of data exchange.

Fragmentation in the smart home, as an exemplar of the collected world, exists across several intersecting axes. Technological fragmentation is paramount and presently persistent, given the panoply of different devices, smart home infrastructures and communication protocols currently in operation. Chapter 3 highlighted that any given smart home is likely to have numerous data communication pathways that consist of data generated from an array of different devices, each with its own sensorised data generation requirement. Even something as simple as user control of devices is fragmented, as most devices can be controlled by different controllers, such as smartphone apps, fobs, hubs, voice assistants or dedicated controllers. On top of these fragmentations, it should not be forgotten that further complexities arise due the multifarious nature of different homes, where the size and design of the home may have an impact on what data is collected and the insights gained. For example, the data collected from a sensorised open-plan loft could be very different to that from a semi-detached house. A 'one size fits all' data collection model could thus produce very different behavioural insights from different homes.

Chapter 3 also demonstrated that there is a significant number of commercial actors involved, such as smart home systems, platforms and device providers. There is no single dominant player, architecture or model of a smart home. Commercial partnership choices based on data exchange are challenging, because while multiple partnerships with smart home system or device providers are possible, partners can quickly come and go. The Neos model, covered in Chapter 4, is a case in point, where the smart home system provider, Fibraro, pulled out of the UK smart home market, thus affecting Neos' business model. Corporate involvement in the smart home environment can be volatile given the dependency on partners, who are navigating their own environments and working out how to commercialise their own sensor data.

Another common fragmentation characteristic also emerges in business models based on data exchange strategies. Chapter 4 highlighted several smart home business models that are currently developing in the insurance space. There is a common characteristic amongst these partnership models: all partnership actors are still searching for value in the smart home environment, and this is reframing the meaning of ownership in many different contexts. Individual corporate ownership of smart home sensor data, for example, is problematic because it is difficult to retain exclusive control of data in a smart home environment. Technological fragmentation means that exclusive data collection structures are challenging. The prospecting of sensor-based value therefore means being thoughtful in making use of existing customer data holdings. For example, Chapter 4's coverage of the Partnered Intermediary model indicates that existing customer data is a key factor in partnership involvement and model development. All three models have different data requirements, some more significant than others, but all are dependent on collection and exchange of sensorised data.

The availability of smart home sensor data and the ability to collect it for aggregation of comprehensive patterns is also fragmented. The volume of sensor data generated in any given smart home is significant, and it exists in multiple formats: video, audio and multiple types of device metadata. The totality of smart home sensor data, in any given home, is generally not capable of being collected from one source given the sheer number of data collectors. Corporate abilities to retain control of smart home sensor data generation is equally diffuse, as valuable insight from the inference of individual behaviours and patterns is more likely to emerge from aggregations of all data, as opposed to individual activities in relation to specific devices. It is this inability of any one party to retain control over smart home data collection environments that is fuelling the development of smart home data exchange partnerships, such as those of insurers and smart home systems. Unless an entity has the capacity to create its own data collection infrastructure that can specifically meet its own purposes, such as the Partnered Data Acquisition model of Neos, then it is required to exchange data as a connective component that generates value, as exemplified in the Partnered Intermediary model.

All this fragmentation unfolds in legally uncertain spaces, at least from an information privacy perspective. Smart home governmental policy is at best tangential, and the smart home is used to fulfil ancillary policy ambitions. The home is traditionally a legally protected space, but the smart home has yet to garner specific legislative, judicial or regulatory attention. As indicated previously, the different regulatory modes of information privacy law that emanate across the globe will likely categorise smart home sensor data differently. Fragmentation will also exist at the macro policy level as well as on the ground as the gradual effect of different jurisdictional analysis unfolds.

The situation becomes even more fragmented when the privacy considerations of individual users are contemplated. Several studies highlight that individual smart home users have different privacy requirements that can change depending on the type of data that is being collected.[43] A smart home user with generally low privacy requirements for sensor collections can nonetheless have higher privacy requirements in relation to home video data.[44] Differentiating between customers with different levels of privacy requirements will be a significant task and re-emphasises the nuances of being involved in the smart home environment.[45] These nuances are further compounded by the social and power dynamics within the home, in which smart home

---

[43] Zeng, Mare and Roesner; Birchley and others.
[44] Zeng, Mare and Roesner.
[45] Wilson, Hargreaves and Hauxwell-Baldwin, 'Smart Homes and Their Users'.

products can be used for purposes such as behavioural surveillance of one individual by another[46] which are simply not contemplated by the data collector.

As such, the smart home is not just fragmented; it is also a contested environment. There is a power dynamic regarding sensorised data collections that is explicitly recognised by some of the information privacy law concepts highlighted in Chapter 5 but is often tacitly disregarded in the smart home literature, as discussed earlier. The tension, of course, regards smart home product manufacturers, as data collectors and consumers of smart home products, as data subjects. The tension is articulated as follows:

the former want to collect as much information as possible for future reference or business opportunity, while consumers primarily only want to reveal small pieces of information that is useful at a certain point in time.[47]

The tension between collector and provider is often masked by the embedded nature of sensorised collections, given their invisibility.[48] However, the importance of highlighting smart home sensorised collections in the privacy context regards the moral component of home life.[49] Sensor-generated collections in the home therefore have a moral connotation.[50] Sensorised collections and environmental measurement appear to be technologically neutral in their collection purposes, but they actually capture a moral insight of home life and have moralistic outcomes. For example, capturing data on home energy usage identifies categories of consumption, and in some cases, over-consumption.[51] The result for the individual, as outlined earlier, is the faintly articulated 'unbidden sharing', but it also means that a corporate data collector will gain an insight into a moral aspect of an individual's or a family's life that the collector does not necessarily want but may nonetheless be required for report or action.

The smart home, as any home, is a contested space that involves several different relationship dynamics. Chapter 3 noted that the purpose of smart home sensor data collection is to identify the activities, behaviours and patterns of individuals in the home. Chapter 4 then highlighted the analysis of smart home insurance partnership models, revealing that the accumulation of sensorised data from the smart

---

46 Zeng, Mare and Roesner.
47 Jacobsson and Davidsson.
48 Zeng, Mare and Roesner.
49 Strain.
50 Lewis.
51 Ibid.

home, and its participants', is the core focus of business model development.[52] Sensor data collected for the purposes of smart home business models is about participant activities in the home rather than the home itself. This point often gets bypassed because the eventual analytic output of predictive or prescriptive processes is purportedly aimed at the functioning of the home and in doing so assisting its inhabitants to live more comfortable lives. Sensorised collections appear morally agnostic, as sensors merely sense and record environmental conditions, including the behavioural patterns of smart home participants. Behavioural monitoring has thus a by-product appearance which re-emphasises that smart home sensor data is the asset to generate value.

However, the first part of the book demonstrated that the monitoring of the individual and the sensorised object is inseparable. In fact, it is the *participant*, rather than the data, that is the 'asset' which generates value. Chapter 4 highlighted that the future value of smart home sensor data, for smart home insurance purposes, is conceptualised in the same manner as telematics data for car insurance. Behavioural monitoring is necessary to better ascertain risky behaviours that can be 'nudged' through real-time, premium-focused risk assessments. Value from smart home sensor data emanates through its ability to finely monitor participant behaviours. Smart home sensor data is therefore not the asset that secures value; instead, the individual is the asset, because the valued smart home sensor data is not generated without the individual undertaking regular, day-to-day activities in the home.

Value in smart home sensor data is derived from structures of subordinated data generation aimed at tracking the behaviours of those individuals that generate data. Commercial value from sensor data is therefore not solely about data generation per se. Rather, prospective value is about the accumulation and analysis of smart home sensor data to gain insight about individual home activities. Smart home sensor data value is about the ability to foster meaningful insight. In the smart home context, as in the broader collected world context, the ability to generate meaningful insight is derived from boundary-reduced spaces, where the reduction of boundaries creates enhanced opportunities for monitoring and tracking individual behaviours.[53] These boundary-reduced environments also give rise to significant challenges for the principled processes and conceptual basis of information privacy law.

---

[52] Austin, 'Re-reading Westin' 55.
[53] Steeves 329.

7.5   Control in Boundary-Dispersed Environments

All the foregoing considerations highlight challenges to information privacy law arising from the increasing ubiquity of sensor data collections. The final challenge, the increasing disruption of control in boundary-dispersed environments such as the smart home, leads to the critical questioning of whether information privacy law can realistically provide the types of control mechanisms aimed at protecting informational spaces for autonomous growth. Boundary dispersion in the smart home occurs in several ways, as highlighted in Part I of the book. Chapter 3 introduced the types of telematic nudging processes that are now emanating in smart home environments. As noted previously, these nudging processes diminish the boundaries of autonomous human control by relocating the information basis of decision making from the individual to the system. Similarly, Chapter 3's anatomy of the smart home demonstrated that it is a highly complex and fragmented environment that involves multiple data exchange partnerships, as evident in Chapter 4's conceptualisation of business models. Control at all levels is, in effect, dispersed differently among data-providing, -collecting and -exchanging parties. Again, sensorised collections are a key factor regarding the emergence of the boundary-dispersed environments of the collected world.

Sensing environments simply do not respect traditional boundary markers and significantly weaken traditional social, moral and environmental boundaries, such as those that typify the home as an essential private space.[54] Physical and spatial boundaries are simply bypassed in sensorised environments, where sensor data can be aggregated to provide a complete picture of all activities. Internal walls as boundary markers are no longer separations of space and private activities because the physical nature of the wall does not prohibit sensorised collections. For example, Wi-Fi signal strength tracking of mobile devices enables fine-grained location monitoring of home participants regardless of physical boundaries.[55]

Informational boundaries therefore become confused. As highlighted earlier, in the analysis of identified privacy concerns, the dispersion of

---

[54] Bryce Newell, 'Privacy and Surveillance in the Streets' in Bryce Clayton Newell, Tjerk Timan and Bert-Jaap Koops (eds), *Surveillance, Privacy and Public Space* (Routledge 2019).

[55] J Zhang and others, 'WiFi-ID: Human Identification Using WiFi Signal' (2016 International Conference on Distributed Computing in Sensor Systems (DCOSS), 26–28 May 2016).

informational boundaries is often encapsulated in the privacy concern of 'the unbidden sharing of data'. The multiplicity of sensorised devices, communication pathways and data collection processes in the smart home means that it is challenging for smart home participants to truly understand the context of data collection. The data analytic quest exacerbates this situation, because the search for the unintuitive demands constant mining of new insight. Maintaining informational boundaries around the notion of individual control, especially through mechanisms of consent, becomes challenging in sensorised environments. In other words, and as covered in more depth in Chapter 8, in relation to Solove's criticism of privacy self-management, the individual simply cannot understand the true complexity of collection, aggregation and output frameworks – especially when those frameworks demand structures of sensor embeddedness to capture truly accurate data representation of natural activities and behaviours.

At the same time, moral boundaries become increasingly blurred, because the smart home is not the private space of the traditional home. As noted above, sensorised data collections on home energy usage reveal consumption patterns which have moral and societal implications.[56] The simple act of locating a sensorised smart home device can therefore have a significant moral connotation.

The effect of sensorised monitoring also means that legal boundaries are challenged. The book has noted several times that legal protections have historically been accorded to environmental spaces, individual activities and types of information based on their essential private nature. The dispersion of physical, informational and moral boundaries also has a concomitant legal effect. For example, the smart home is no longer the private setting of the traditional home, and thus how legal protections apply in this realm becomes increasingly unclear. The foregoing analysis highlights that the threshold question of whether smart home sensor data is personal information is a complex issue that does not provide a straightforward answer when considered across the legal requirements of different jurisdictions. The scope of information privacy law protections is therefore unclear in collected spaces like the smart home, because technological frameworks are becoming more fragmented and informational boundaries are becoming less defined.

The dispersion of boundaries has a corollary effect on corporations seeking value from sensor data in smart home spaces. The new searches

[56] Lewis, 'Energy in the Smart Home'.

for commercial value, outlined in Part I, undermine corporate data collector ability to retain control of sensor data in environments where all forms of boundaries have been dispersed. Sensorised data collection cannot therefore be constrained or utilised individually by one corporate smart home data collector. Value requires meaningful insight, which in turn requires aggregated sensorised collections involving data beyond the control of any individual smart home data collector. A counter-intentional consequence thus arises: smart home control requires more fine-grained data on individual activities and patterns to provide meaningful insight. However, meaningful insight resides outside of the control of an individual corporate data collector, as exemplified by the Partnered Intermediary and Platform Entity models conceptualised in Chapter 4. An important point thus arises for corporate data collectors that impacts upon notions of individual control processes in information privacy law – namely, the search for commercial value in the smart home may not equate to the desire for control of smart home sensor data.

To secure value in this boundary-dispersed environment, corporations may need to give up their desire for control, and ultimately the desire for ownership of the asset that enables value generation. If the valued asset has indeed changed from the data to the participant, it becomes challenging for corporations to seek control and ownership of the asset. The participant increasingly has greater degrees of control because more and more services involve the participant through subordinated data collections in sensorised home environments. The participant therefore must be meaningfully involved in the collection process; otherwise it is not possible to derive valid insight from smart home sensor data. Delegated smart home data collections mean that the focus of meaningful insight, as a way of generating value, requires a shift in corporate perception of the customer and their relationship to the business entity.

The foregoing analysis demonstrates that there are changing boundaries at all levels in the smart home. The traditional home is encapsulated by the ability to build boundaries and to make it private from the public realm. It was conceptually constructed on the affordance of boundaries which become less bounded in the smart home setting. The smart home simply seeps sensor data that disperses through the physical, spatial, informational, moral and legal boundaries of the home. The smart home thus disperses the very notion of boundary and minimises the ability of all parties involved in data collection to create, maintain and manage boundaries. In effect, the involvement of multiple corporate actors and technological frameworks shapes the many different relationships involved in the smart home and reconfigures relationship boundaries between home participants themselves, home participants and smart home service providers, and service provider and service provider.

These are challenging issues for any corporate data collector of which compliance with the information privacy principles identified in Chapter 5 will only assist so far in the fragmented, contested and boundary-dispersed environments of the collected world. For example, the principle of data minimisation is challenging. Ubiquitous collections of smart home sensor data are required for the smart home to be smart. Ubiquity, by its very nature, challenges the notion of minimality and the basis that smart home data collectors should only collect sensor data to fulfil a specific and identifiable purpose. The uncertain search for value in the smart home environment means that sensor data collection must be prospective.[57] The continuous collection of data also means that new behavioural insights are always likely to emerge. Data collection strategies are therefore circular. More sensors mean more data. More data means more insight. More insight means more prospects of value.[58] Data minimisation may operate effectively in binary processes of personal information collection, but it is uncertain as to how the principle would operate in circular and multiple collection processes of always-on sensor data.[59]

Similarly, the purpose specification and disclosure limitation principles seek to ensure that personal information is only used for specified purposes and can generally only be disclosed for those purposes. However, sensorised collections will enable the identification of many different behavioural patterns that will go beyond the purpose of collection and use. Telematics, a case in point, is where sensorised collections of driver behaviour can also be used to infer the states of mental well-being of drivers, as detailed in Chapter 3. In the dense sensor environments of the collected world, the regulatory boundary between primary purposes and secondary uses also becomes dispersed. As noted earlier, the environmental capture of sensor-derived monitoring will enable insight of all aspects of life, as indicated by the smart home analysis. The distinction between a secondary use and a primary purpose becomes blurred in the context of environmentally focused collections that inherently intertwine the recording of one activity with another. The monitored

---

[57] MacCarthy 57 as part of the defence of Big Data and the reduction of ex ante secondary use restrictions to a risk-based system.

[58] Andrejevic and Burdon 32.

[59] van Hoboken 248 in the context of Big Data referring to the use oil 'on the slippery slope of data collection and unpredictable use'.

use of utilities in the smart home is thus inseparable from the identification of individual moral decisions about that use.

Chapter 3 and this chapter have detailed the significant difficulties of securing personal information in a smart home context. In such a fragmented system, the very simple question of who bears security responsibilities and obligations becomes a challenging one, because the boundaries of regulatory security allocation are dispersed across multiple actors who are required to respond in multiple ways. Attempting to quantify reasonable security measures in infrastructures that house devices with inbuilt security limitations becomes problematic. Equally problematic are the principles of information quality and individual participation. Again, the expected bounds of individual rationality are extended to their limits. For example, it is one thing to expect an individual to be able access collected data to check the accuracy of individually provided personal information such as date of birth and other readily checkable forms of data. However, it is a completely different situation where an individual will be required to check the accuracy of sensorised data collections to confirm that any given device sensor is recording valid and reliable data about behavioural patterns at any given time in which the device is working and collecting data.

7.6   Conclusion

The challenges outlined in this chapter give rise to some notable tensions that are covered in the remaining chapters of the book. These tensions are fundamental and cannot be avoided by corporate data collectors in collected environments like the smart home. The tensions are systemic and built in. In effect, these tensions are the system. The tensions reflect the fact that the system is built for, and by, multiple actors through multiple technological frameworks and seek to service yet-undefined value propositions. The fundamental tensions that arise therefore give rise to some significant challenges for information privacy law that need to be thoughtfully considered. Section 7.5 highlights the development of the smart home as a boundary-dispersed environment. Nevertheless, the desire for boundaries remains, at all levels, to retain control of different aspects of the smart home environment.

Home participants still have expectations of privacy in the home and seek to make that space private, even though the introduction of smart home infrastructures has the opposite effect. Equally, smart home system and service providers seek to establish a commercial footing through smart home sensor data, even though they cannot control value generation. Similarly, traditional home-focused services, such as home insurers, seek to maintain commercial relevance in the smart home, even though technological frameworks require them to have less control over ownership of customers and customer data.

There is a fundamental tension, therefore, between the desire of all parties to retain control in the smart home and all parties losing control over the ability to build boundaries in collected environments such as the smart home. Linked to the first tension is the nature of the smart home itself and what makes the smart home smart. There is a further fundamental tension:

1. For the smart home to be smart, it needs ubiquitous collections of sensorised smart home event data that can be used to monitor and predict participant behaviours.
2. However, ubiquitous collections reduce participant understanding of the home and thus reduce participant knowledge and ability to control internal and external boundaries. As pointed out previously, it may not matter whether a participant closes the curtains to shield what's happening inside a home because sensorised environments can bypass physical boundaries and inform of activities through the accumulation and analysis of smart home sensor data.

The smart home may get smarter through ubiquitous smart home sensor data analysis, but the human participant may become increasingly less knowledgeable about the monitoring of behavioural activities in the home. This could have significant consequences for the application of information privacy law frameworks based on the notions of control, privacy principles and informed individual participation, particularly through the process of notice and consent.

The lack of participant knowledge also gives rise to a fundamental tension regarding participant decision making in the smart home. The smart home facilitates increasingly automated decision-making processes predicated on the prediction of participant behaviours, such as in the Partnered Data Acquisition model and through smart home monitoring and prevention services, highlighted in Chapter 4. However, automated decision-making processes further reduce the autonomy of individual participants, especially where an algorithmic framework of devices and services, such as Nest, automatically adjusts the home.

Smart home infrastructures thus increasingly become home decision makers and either make adjustment decisions automatically or can nudge participants towards a certain outcome. A very different type of decision making consequently pervades. One of the consequences of a boundary-dispersed home environment, therefore, is potentially the diminution of individual autonomy in making home-based decisions.

An autonomy deficit could have a detrimental impact upon a participant's ability to understand the basis of automated decisions and the collection processes involved. This gives rise to a fundamental tension involving control:

1. As detailed in this chapter, all parties in the smart home want to retain individual control of the environment.
2. However, control is no long individually focused; instead, control is *systemically dispersed.*

The dispersion of boundaries means that the ability to control now seeps through the system to the extent that it becomes challenging to address the fundamental question of *who is in control?* The dispersal of boundaries means control is a diffuse concept in the collected world. No individual party can claim outright control even though they all desire to do so. All of these factors lead to a fundamental tension entailing information privacy:

1. Privacy in the collected ecosystems like that of the smart home is largely conceptualised as information privacy, predicated on the ability of individual smart home participants to retain control over their personal information.
2. However, the removal of boundaries reduces the ability of any one part of the ecosystem to control any data, including a participant and their control over personal information.

If smart home participants increasingly do not understand the complexity of ubiquitous smart home event data collections, they cannot respond meaningfully as autonomous individuals seeking control over personal information. Similarly, smart home systems and services need aggregated smart home sensor data to generate meaningful insight, but one system or service cannot get access to all data, even though they are an intrinsic part of the collection process, due to the fragmented nature of the environment. All these tensions reflect the fragmented nature of the smart home environment. It should also be borne in mind that fragmentation exists at all levels: technological, social, commercial and legal. The home is a complex environment, and the smart home is even more so.

This chapter has outlined the challenges to information privacy law processes, principally in application but also in concept. A question therefore arises: should we simply give up on information privacy law, given the scope of the challenges presented in one aspect of the collected world – namely, the smart home? If the law cannot offer protections in one of the most protected liberal spaces, then what chance does it have of being able to provide protections in less protected spaces and environments? These questions are explored in the next chapters from a conceptual and application standpoint that moves to consider the infrastructural context of information privacy and its role in interrupting modulated forms of power.

8   Conceptualising the Collected

8.1   Introduction

The remaining chapters of the book consider the distance travelled so far, and they articulate some conceptual and legal responses to the collected challenges outlined in Chapter 7. Legal responses focus on the application of current information privacy law structures and seek to shift information privacy from a control-oriented mechanism to something different. The tacit theme that flows throughout the book is whether the protections accorded by information privacy law are adequate in the dawning of this new collected reality brought about by the ubiquity of sensorisation. Critical attention thus returns to the concepts and legal processes of information privacy law introduced in Chapters 5 and 6. In so doing, the question shifts from the descriptive to the normative: namely, in the collected world what should information privacy law seek to protect, and how should it provide protections, particularly in relation to types of regulated information and the application of collection principles?

This chapter critiques the dominant conceptual scope of information privacy law frameworks that overly focus on individual control mechanisms over personal information. Following on from the challenges outlined in Chapter 7, it will be argued that information privacy law should be considered in a conceptually different way. Building on Julie Cohen's cumulative work of the past decade,[1] it will be argued that the conceptual basis of information privacy law should encompass the relational and societal context of information exchange and consider more specifically the power-related implications that flow from it. Utilising the coverage of the smart home and smart home insurance

---

[1] Cohen, *Configuring the Networked Self*; Cohen, 'What Privacy Is For'; Cohen, 'Between Truth and Power'; Cohen, 'The Networked Self in the Modulated Society'; Cohen, 'The Surveillance-Innovation Complex'; Cohen, 'Affording Fundamental Rights'; Cohen, 'The Biopolitical Public Domain'; Cohen, 'Turning Privacy Inside Out'; Cohen, 'Review of Zuboff, Shoshana. 2019'.

models, the considerations in this chapter will move from the micro-positioning of data collection in the smart home to examine the infra-structural and systemic issues that arise from data collections based on sensor ubiquity. Some of these broader issues emanate specifically from the Platform Entity model, detailed in Chapter 4, which provides a background through which to examine the attempted concentrations of platform power over sensor data collection from environments such as the smart home. All of this leads to issues of modulation, as Cohen directs us to consider.

In relation to the critique of information privacy law's control basis, five intended outcomes of the law in application are adduced – namely, the enhancement of autonomy, the amelioration of power imbalances, the mode of transactional operation, the use of information disclosure mechanisms to ameliorate informational asymmetries, and the over-arching balancing mechanism. The challenges highlighted in Chapter 7 are then applied to these intended outcomes to show the limits of the control model, particularly in relation to new understandings of power. It is argued that the conceptual basis of information privacy needs to consider more specifically the relational and societal implications of data collection and exchange in a collected world. Put simply, the col-lected world fundamentally challenges the predicate base of privacy self-management as the foundation of information privacy, as noted by Solove.[2]

A fundamental part of reconsidering information privacy is the explicit acknowledgement of the role of power in shaping the infrastruc-tural understanding of the collected world. To better understand the emerging forms of power play that emanate from a collected state of being, and the effects of that state on information privacy law, Cohen's core conceptualisations of a modulated life – namely, modulation, semantic discontinuity and operational accountability – are used to highlight the structural and hegemonic flows of biopolitical power that fundamentally challenge the basis and application of information privacy law. Several movements from the five intended outcomes of the control model are therefore required: from autonomy to situated intersubjectiv-ity, from power vacuums to modulation, from transactional operation to boundary management, from information asymmetries to social shap-ing and, finally, from balancing mechanisms to exposing modulation.

A different conceptual form of information privacy thus emerges – one that is more suited to the challenges of the collected world.

---

[2] Solove, 'Privacy Self-Management and the Consent Dilemma'.

8.2   Moving from Control

The spread of information privacy laws across the globe continues to grow. There are now over one hundred countries that have some form of information privacy or data protection law largely based on the same type of principled protection that emanated from the three founding legal instruments detailed in Chapter 6. These laws, as noted in Chapter 5, are based on mechanisms of individual control as an intrinsic element of balancing different interests. A question therefore arises: why question the established basis of the conceptual and legal application of information privacy law as an individual, control-based processual mechanism? This is a reasonable question to ask given the seemingly successful implementation of information privacy laws throughout the world. To answer that question, it is helpful to ask two further questions and to return to the preceding coverage of this book, particularly the collected challenges. The two further questions are these: what are the *intended outcomes* of a control-based information privacy law? What does control-based information privacy *do*? The two further questions hark to the control-based purposes of information privacy and the effect of those purposes in practice. Answering these two questions therefore assists in addressing the primary issue of why to critique the control basis of information privacy in the first place.

Let's start with the first question and look at what the control basis of information privacy is intended to do. Chapter 5's coverage noted the overlapping conceptual themes of information privacy. The same can be said of the conceptual purpose of individual control, because it entails of information privacy law, for it to be effective, to fulfil certain underlying predicate requirements. Like the broader themes, these requirements overlap and are interlinked with each other to provide a holistic base of protections. Some requirements are broad in focus and regard the relationship between privacy and the protection of individual autonomy in a democratic and political sense. Some requirements are narrower and seek to provide individually focused process mechanisms to support the broader societal and political requirements of control. However, much like the operation of the processes of principled protection of information privacy law, the requirements are intended to operate together to meet the different intentions of control-based application. All these requirements have been introduced in different parts in the book, and this section now begins the process of cumulative analysis and that of tying together the different threads of the book to derive a reformulated narrative of information privacy necessary to meet the collected challenges outlined earlier on.

At its broadest level, the control basis of information privacy seeks to respect and augment the autonomous capacities of individuals regarding exchanges of personal information. As highlighted in Chapter 5, however, it does so in a tacit fashion. The reasons for its tacit nature are complex and, again, are often interlinked with other characteristic developments that have taken place since the concept's inception. The explicit focus on control mechanisms, as opposed to broader autonomy purposes, is probably a reflection of the timing of the control concept's genesis in the 1960s and 1970s and their development in conjunction with the beginning forms of mass data collection and computing. Many of the originating concerns highlighted by early proponents, such as Alan Westin[3] and Arthur Miller,[4] were specifically related to this issue. It is not perhaps surprising therefore that the conceptual focus of control has been constructed around technocratic mechanisms of process-based forms that mirrored the advent of newly developing administrative processes of personal information exchange.[5]

Another key factor in the tacit considerations of autonomy involves the timing of the first information privacy laws. The first iteration of US FIPPs was introduced within a few years of Westin's *Privacy and Freedom*, and the first European data protection laws started to appear at the same time. Within a decade leading up to the early 1980s, the three founding instruments and the regulatory tracks they laid had already set the foundation for the future development of information privacy law based upon technical and administratively focused processes of information exchange. These developments also took place at a time when there was great uncertainty in the legal academy about whether it was even possible to adequately define privacy in a meaningfully legal sense. The control basis of information privacy laws and their readily identifiable process protections made the broader conceptual discussions about the legal meaning of privacy, including the role of autonomy that privacy placated, largely redundant. Control-based information privacy, whilst seeking to protect autonomy, thus focused on identifiable mechanisms of process interaction that could manifest with relative ease in legal and regulatory frameworks.

Chapter 5 also highlighted that the tacit nature of control-based autonomy considerations implicates deeply embedded concerns about the imbalance of power relations between data collectors and individual personal information providers. Again, the power-related concerns

---

[3] Westin.
[4] Miller.
[5] Gonzâlez Fuster.

were largely constructed around increasing governmental and private sector data-collecting and computing capabilities, as noted in Solove's much-cited work on metaphors of power in information privacy.[6] Solove makes clear that the control basis of information privacy, through process-based protections, is also intended to ameliorate the power imbalances between data collector and provider. The points of individual interaction required in the control-basis application of information privacy's life-cycle protections were designed to create autonomous spaces for critical reflection that manifests in meaningful and informed decision making. Interaction spaces are therefore small points of power vacuum that provide control levers for individuals regarding exchanges of personal information with entities that are greater in size, scale and scope. The augmentation of individual autonomy and its purpose in reducing power imbalances are therefore intrinsically enmeshed within the notion of control.

These broader requirements of control are further supplemented by narrower ones. However, while the narrower requirements are closer to on-ground implementation considerations, they nonetheless have an impact upon the broader foci of control that manifest in the three jurisdictions detailed in the book. In one sense, the three narrower requirements shape the jurisdictional focus of legislative and regulatory impetus. Those requirements – namely, modes of transactional operation, information disclosure measures to combat information asymmetries and balancing mechanisms – are, as with other aspects of information privacy, interlinked and impact upon each other in application.

At the heart of information privacy's control basis are modes of transactional operation in which the exchange of personal information is essentially viewed as a tradable exchange.[7] In other words, the provision of individual personal information is undertaken in conjunction with a trade of data for a good or a service. Chapter 5 highlighted the links between the control basis of information privacy and constructs of property ownership in personal information. Chapter 6 then noted that these constructs resonate more strongly in the market-based, sectoral regimes of the United States than in the rights-based perspective of the EU. Chapter 4's coverage of the emerging data exchange partnerships of smart home insurance re-emphasised the economically transactional trading nature of personal information exchange in the United States.

---

[6] Solove, 'Privacy and Power: Computer Databases and Metaphors for Information Privacy'.

[7] Schwartz and Peifer 147 particularly in relation to US laws.

Both the Partnered Intermediary and the Platform Entity models specifically provide trading mechanisms for individuals to transfer certain elements of smart home sensor data to insurers for a better-priced premium. Trade of personal information is central to the operation of these models and the application of information privacy law apparatus, in the form of privacy policies that flow from them. The Partnered Data Acquisition model is also based on the notion of trading sensor data. However, unlike the other two models, the point of individual decision making about trade is required at implementation of sensorised devices, as part of an agreed serviceable trading arrangement for better-priced insurance premiums based on the monitoring of behaviours and patterns. The actuality of personal information exchange can therefore also result in data trades under the auspices of the rights-based framework of the GDPR. However, there is an economic rationality to the US regime of information privacy that is more heavily oriented towards tradability than is the case in other jurisdictions.

It is no surprise, as also noted in Chapter 6, that US information privacy laws place such heavy regard on the process of notice and consent as the primary protective mechanism of its control-based laws. The preference for the notice-and-consent model is directly linked to the economically functional structures of US information privacy law, but it is also an intrinsic feature of all jurisdictions' autonomy-enhancing intentions. The notice-and-consent model of the United States, the consent requirement for legitimate processing under Article 2 of the GDPR in the EU and the application of Privacy Principle 1 in Australia all regard the provision of privacy policies as essential information disclosure mechanisms. These mechanisms fulfil the broader requirements of control-based information privacy, as they assist in providing informational spaces for autonomous decision making and seek to rebalance the power aspects alive in information asymmetries.

Privacy policies, as information disclosure mechanisms, also impact upon legislative and regulatory frameworks that orient towards personal information as a tradable commodity or asset. In other words, the use of privacy policies, such as the fulfilment of notice-and-consent obligations in the United States, reflects and shapes the underlying regulatory perspective of control. The market-based structure of the US laws focuses more heavily on the information disclosure aspect of privacy policies that are intended to emanate in a decision about trading personal information. As noted in Chapter 6, there is a direct link in laws such as the GLBA between the purpose of corporate information disclosure and a subsequent tradable decision by an individual. However, in the EU, the privacy policy does not engender the same economically rational, information disclosure intentions. Instead, the privacy policy and the GDPR's use of consent as a marker for legitimate processing provide foundations for fair information exchange, where fairness is deemed an essential and functional constituent of data protection rights. To a lesser extent, the same is true for Australia, except that the role of fairness is used as a balancing yardstick rather than a rights-based signal of intention.

All of this brings us to the final, narrower requirement that binds up all the considerations covered thus far in this section. The control basis of information privacy is about different degrees of balance and the different mechanisms employed to secure balance between individual providers and organisational collectors of personal information. There is an explicit acknowledgement in the control basis of information privacy that the different interests of the provider and the collector are at play. The control basis of information privacy therefore provides mechanisms to balance and preserve the interests of individuals, collectives and societies. The balancing requirement is most clearly expressed in the original OECD Guidelines, as highlighted in Chapter 6. The balance requirement is considered so important that it is represented legislatively in the Australian Privacy Act and in other OECD-country legislation that has used the Guidelines as a basis for implementation. For example, Section 29 of the Act specifically requires the Privacy Commissioner to pay due heed to the Act's objects, which recognise, at Section 2A(b), 'that the protection of the privacy of individuals is balanced with the interests of entities in carrying out their functions or activities'.[8]

The five intended outcomes of effective control-based information privacy law highlight many of the conceptual and practical complexities highlighted in Part II. However, the collected challenges detailed in Chapter 7 make it difficult for these intended outcomes to have the anticipated material effects in delivery. Those challenges – namely, the predominance of sensor data collections from fragmented and contested environments – disperses any party's ability to control data collection and exchange practices. The control basis of information privacy is therefore challenged in the sensorised, fragmented and contested environments of the collected world.

---

[8] See, for example, s3 Personal Information Protection and Electronic Documents Act S.C. 2000, c.5, which requires a balance between 'the right of privacy of individuals with respect to their personal information and the need of organizations to collect, use or disclose personal information for purposes that a reasonable person would consider appropriate in the circumstances'.

Chapter 7 noted that sensor data collections are key to smart home business model developments. It is also important to bear in mind, from Chapter 2's coverage, that the focus of sensorised collection is not just about the home. The book has focused on the smart home as the paragon exemplar of sensorisation in the collected world. It is the cherished space of liberalism due of its role in encouraging the autonomous growth of individuals. Chapter 2 highlighted the interconnected use of sensor data across three levels of smartness: the smart individual, the smart building and the smart environment. Chapter 7 also noted that the collection and use of sensor data has a different effect than that of traditional forms of personally identifiable information.

There is a deeper complexity about the nature of sensor data that involves its relative ease of collection and its ultimate scale of use. Various parts of the book have highlighted the use of sensor data to identify, at ever-finer degrees, individual patterns and behaviours that can be collected at scale through a range of different sensorised devices. Consequently, the focus of data analytic processes involving sensor data can simultaneously switch focus from individuals to communities and to populations with ease. Monitoring one, many and all becomes seamless, as detailed in Chapter 2's coverage of mobile phone sensor data (which has individual, building and infrastructural application). The very same sensor can generate data that can be used to identify when a sufferer of Parkinson's disease is having an attack or to identify earthquake tremors that have a much broader population and societal impact.

Sensor data collections and uses are indeed different to the types of data collection and use envisaged by the control-basis model of information privacy law. That model, as highlighted previously, is process based, to mirror administrative and technical personal information exchange life cycles. However, as Chapter 7 points out, the circular, continuous, multiple and multifarious collections of sensor data that now pervade as the norm bear no resemblance to the binary, one-off, singular and purposeful data collections of the past. The fundamental characteristics of sensor data collections make it more difficult to provide mechanisms of individual control, especially where it is not even clear whether generated sensor data would give rise to suitable classification of protected information that would engender information privacy law protections. Chapters 4 and 5 demonstrated the different constructions of sensor data as either personal data under the GDPR, as not PII under the US sectoral frameworks and possibly or possibly not personal information under the Australian Privacy Act. As such, the notion of control redolent in individual autonomous decision making, meaningfully informed by privacy policies as part of notice-and-consent models, also becomes problematic.

Control problems also emerge in data collection spaces that are increasingly fragmented and contested. The technological framework of the sensorised environments that make up the collected world are complex. The smart home anatomy, detailed in Chapter 3, showed that many data collectors seek to collect data from a proliferation of different devices and communication frameworks. Different controllers can be used to control a plethora of smart home devices that leave data trails which are collected by alternative communication pathways. These pathways extend beyond the home and into cloud-based platforms that can reside in different jurisdictions across the globe. It is simply not clear what the implications are of using a smartphone app as a device controller compared to a dedicated fob or a voice activation digital home assistant. The complexity of these multiple data collections and exchanges therefore makes it challenging for any individual to fully comprehend their own environment and to make the type of informed, meaningful and autonomous decisions required by the control basis of information privacy law.

Chapter 7 also highlighted that environmental sensorisation and technological fragmentation cause a dispersion of informational boundaries in the smart spaces of the collected world. This dispersion greatly reduces the capacity for individual control over the use of smart spaces, including sensor data collections that are intended to identify individual behaviours in conjunction with environmental operations. The traditional home is encapsulated by the ability of its inhabitants to build boundaries and to make the home private from the public realm. Physical boundaries, such as locks on doors or drawn curtains, allow individual inhabitants to separate the home from the outside world to create spaces of private solitude. These spaces are also possible in the home, where certain rooms – such as bedrooms, bathrooms and toilets – can be separated from other ones and their individual inhabitants. Control over a physical boundary can thus be conducted with relative ease. A door can be closed, and a key can be turned. However, as noted in Chapter 7, the permeability of wireless collections of sensor data from always-on devices can bypass the physical boundary. In other words, control over a locked door is no longer a privacy boundary, because activities can still be monitored within that space if it is sensorised and sensor data is being collected.

Control over informational and moral boundaries also becomes challenging. As noted in Chapter 5's coverage of Westin's limited and protected communications, part of the established claim of control-based information privacy is the ability of individuals to withdraw from the provision of certain information that would be too sensitive or private to reveal. The risk assessments behind those decisions are possible when they involve identified types of data that will be used for specified purposes – for example, where an individual declines to provide a prospective employer details of potentially discreditable aspects of their life in case it has a negative impact upon future employment prospects.[9] It is possible to make that decision because the consequence is relatively quantifiable. There is an individual and shared social understanding of how the information[10] is likely to be perceived, which informs a decision to withdraw it from the public gaze.

The same consequential calculus is more problematic with continuous collections of sensor data that are used for behavioural identification based on our most natural states of being. The embedded nature of sensor data collections is such that we are generally required to be kept unaware of the sensor and its data collection. The state of sensor unawareness inhibits the development of individual and shared social understandings of when and how to withdraw information from the broader public, or in this case corporate data collectors. The means of withdrawal are also challenging in sensorised environments, as withdrawal either requires a conscious change of unconscious individual patterns of behaviour, over greater periods of time, or to nullify the sensor's data collection capabilities by turning off the device. The former withdrawal mechanism is challenging because an individual is unlikely to know what behaviour to change in data collection environments where the value proposition is equally uncertain. The latter mechanism would re-emphasise the nature of the sensorised object as a 'thing', as it would lack functional utility without its ability to collect and exchange sensor data. A turned-off Google Home would only have utility as a physical object such as a paper weight, albeit an attractive one.

The foregoing analysis questions whether the intended outcomes of control-based information privacy law can be achieved in a collected world. The five intended outcomes of the control model – namely, the enhancement of autonomy, the amelioration of power imbalances, the mode of transactional operation, the use of information disclosure mechanisms to ameliorate informational asymmetries, and the overarching balancing mechanism – are individually and collectively challenged by the characteristics of ubiquitous collection based on sensorisation. Attention consequently shifts to the second question asked earlier, which requires a critical examination of the control model's effect. Solove's criticism of privacy self-management is important in this regard, as the collection of smart home sensor data covered in this book re-emphasises Solove's concerns.

---

[9] Posner.
[10] Austin, 'Re-reading Westin' 73.

As noted previously, including in Chapter 6, the control basis of information privacy manifests in a process of principled protections. It is now argued that the control basis of information privacy overtly focuses on certain points of the process, at the expense of other points. In effect, the control basis promotes certain points over others. Notification as part of an overall effort to enhance autonomy in decision making is a prime point in question, particularly in the notice-and-consent regime of the US. The promotion of notification as a vital strategy has the tendency of promoting disclosure over collection and minimising the role of other principles such as individual participation, quality, access and correction. Individual control is the basis of life-cycle protections, but its aim is not of equal weight. The use of the eponymous privacy policy as an information disclosure measure is paramount in this regard. Individual understanding of a privacy policy is synonymous with a decision relating to the exchange of personal information. As such, individual decisions about personal information provision are really decisions about understanding generated from a privacy policy. In these circumstances, it is questionable whether the control basis of information privacy can provide the type of protections to enable rational individual decisions about provision and use of personal information, as highlighted in Solove's critique of privacy self-management.

Privacy self-management places a significant decision-making burden on individuals to be rationally responsible for the consequences of their information privacy actions and decisions. As noted earlier, the control basis of information privacy attempts to utilise the privacy policy as the primary mechanism of information disclosure that forms rational and autonomous decision making about personal information exchange. Solove contends that privacy self-management suffers from two fundamental flaws that regard the limited cognitive capacities of individuals to make rational decisions, especially in the face of structural complexities arising from the increasingly vast scale of data collection and aggregation.[11]

The basic structure of privacy self-management is dependent upon the capabilities of individuals to be adequately informed about the consequences of their decision through a privacy policy. However, this basic structure is problematic because most individuals do not read a privacy policy when deciding to exchange personal information. Individuals are thus largely uninformed when making exchange decisions, which has the effect of skewing cognitive decision-making capacities.[12] Even if an individual does read a privacy policy, it is still not certain whether they have fully understood the consequences of their decision, because

---

[11] Solove, 'Privacy Self-Management and the Consent Dilemma'.
[12] Ibid 1886.

reasons for consenting to a disclosure of personal information are complex. Privacy policies, themselves, take a degree of knowledge to understand.

Take, for example, the privacy policies of the smart home devices examined in Chapter 4. It took a significant degree of legal analysis to identify the simple question of what sensor data would be treated as personal information. The control basis of information privacy requires meaningfully informed decisions from individuals, but the disclosure mechanism of a privacy policy is insufficient to meet the lofty aims of rational decision making. Part of the reason for this informational failure is that a privacy policy cannot match the contextual complexity of information privacy, and it merely repeats the different parts of the exchange process to individual providers. In other words, it does not adequately relate or conceptualise the contextual challenges of personal information exchange.

These cognitive problems are compounded by the structural complexity of information societies that simply do not allow individuals to fully understand them. In effect, the processes of data exchange are becoming too big for any individual to understand.[13] Moreover, as noted previously, the increasing drive towards sensor data collection requires individuals to be generally unaware of the collections and subsequent uses of that data. Accordingly, it becomes challenging to assess the sensitivities around disclosure of personal information. Chapter 3's coverage of the smart home is a case in point, where it is unclear exactly what data is exchanged and to what collector. It therefore is difficult to assess the potential harms of information provision and disclosure. Yet that is what privacy self-management asks individuals to do at the first point of interaction – the collection agreement – even though most perceptible harms will not emerge until they emanate at the other end of the information exchange process through disclosure.[14]

Critiquing the control basis of information privacy at the point of collection also raises significant issues regarding its transactional mode of operation. Control, in this mode, is part of a tradable transaction of personal information, as highlighted previously. A further conceptual skewing arises that leads to personal information being treated as a form of property that can be owned. The skewed effect unfolds in two ways. First, the property foundations of transactional operation fundamentally reduce the rich scope of information privacy law from issues

[13] Ibid 1888.
[14] Ibid 1891.

of power or autonomy to issues of process. However, as Solove notes, the focus on process issues does not assist in addressing the more fundamental questions of personal information exchange.[15]

Second, the control basis of information privacy law reduces profound questions about the role of individuals in liberal societies and their relationship to powerful data-collecting entities. The control basis diminishes issues of autonomy to issues of information disclosure as a process of ameliorating information asymmetries. This constant limiting of scope emerges in the primary source of problem definition of control-based information privacy, which begins, and potentially ends, with classifications of personal information. As such, the control basis does not really provide the conceptual framework to address the deeper issues arising from personal information exchange. In that sense, control is an ideal, but in practice it is an illusory ideal[16] with a limited conceptual basis which has the effect of reducing its scope in application.

The book consequently turns attention to the conceptual underpinning of information privacy that is necessary to meet the collected challenges outlined previously. The cumulative work of Julie Cohen is put forward as a frame to reorder the conceptual base of information privacy, from solely one of individual control to one that is much broader and thus able to encompass key components of the other conceptual themes highlighted in Chapter 5. In so doing, as detailed in Chapter 9, the focus of the book shifts predominantly from the construct of 'collected' to that of 'modulated', to reflect the complexity of Cohen's sophisticated analysis of networked societies, which provides a conceptual torchlight that enables the tacit issues of power in information privacy to be brought into the light.

This chapter's remaining coverage focuses principally on the critique of the control basis of information privacy by highlighting how some of the key elements of Cohen's work are suited conceptually to the collected challenges outlined previously. In that sense, this chapter considers Cohen's work from several different spectrums that all relate to the micro activities of sensorised data collection from the smart home setting. Chapter 9 will put forward several responses related to the application of information privacy and its operation in practice. Chapters 9 and 10 cover Cohen's analysis of modulation; the situated self; the biopolitic, semantic discontinuity; and operational accountability. These complex concepts, when combined, provide a clarion call about the unquestioned expansion of informational capitalism that is dependent

---

[15] Solove, 'Privacy Self-Management and the Consent Dilemma'.
[16] Hartzog.

upon a collected world for its ultimate data collection fulfilment. Hence the importance of Cohen's work in understanding the collected consequences described in this book.

8.3   Moving to Interruptions of Modulation

Section 8.2 addressed reasons why the control basis of information privacy should be questioned because of the collected challenges that are now emerging. It showed that the five intended outcomes of control-based information privacy law will face problems of conceptual and practical application in a collected world ubiquitously populated by sensorised devices. Cohen's work is now introduced for two reasons. First, it is to complete the critique of control-based information privacy by highlighting the shift from the five intended outcomes of individual control to the foundational bases of Cohen's argument. In doing so, it becomes possible to clearly acknowledge the overlaps between the conceptual themes outlined in Chapter 5 and to provide a new understanding of information privacy's role that explicitly regards power relations. Second, Cohen's work provides a broader basis for information privacy that is better suited to the challenges of sensorisation redolent in the collected example of the smart home. In effect, the broader basis of conceptual coverage is better able to identify and acknowledge the power-related issues that emerge from collected spaces as a propounding of modulation.

Before we get that far, it is worth revisiting Cohen's notion of privacy and thus information privacy, as it sets a frame for understanding the subsequent analysis. Chapter 5 highlighted that Cohen's definition of information privacy is inherently relational, as it enables individuals both to maintain relational ties with others and to develop critical perspectives on the world around them.[17] Those critical perspectives are vital because they are a constituent part of self-development. The role of privacy therefore relates to boundary management that creates breathing spaces in socially situated processes.[18] Privacy is consequently a dynamic condition that creates spaces of play in which persons undertake activities of self-definition and understanding.[19] As discussed in the remainder of the book, Cohen's notion of 'play' is important, because it is a means of restricting processes of power that seek to shape

---

[17] Cohen, 'What Privacy Is For' 1906.
[18] Ibid 1911.
[19] Cohen, 'Turning Privacy Inside Out' 12.

self-making activities. The role of technology in networked societies is also important in this regard, because technologies, such as sensorised devices, have been constructed both to shape understanding and as means to shape behaviour. In that sense, the flow of information from sensorised devices operates as a two-way flow that informs about individual actions and provides the basis for shaping future actions.[20]

Privacy is therefore about boundaries and boundary crossings.[21] Privacy infringements emerge through boundary violations that effect spaces of playful development. However, because privacy is dynamic and socially situated, privacy 'expectations and behaviours are unruly and heterogeneous', which means that they are difficult to reduce to a core set of conceptual criteria.[22] Moreover, privacy expectations are not just informational, because they also emerge through relational, contextual and spatial characteristics.[23] A primary threat to privacy in networked societies emerges through newly forming infrastructures that have the intended or unintended effect of creating structures of systematic surveillance.[24] The emerging surveillance structures of informational capitalism effectively transcend spatial, contextual and informational boundaries to enable ubiquity in data collections. It is this ubiquity, in conjunction with data analytic techniques, that enables categorisation of populations to become targets for the types of commercially driven, predicted and prescripted outcomes highlighted in Chapter 3.

The coverage of Cohen's consideration of privacy is immediately different to that of the control basis of information privacy. Information privacy is inhered within a broader frame of privacy in which information, space and social context are equally melded. In that sense alone, information privacy cannot solely be about retentions of individual control in information exchange processes, because the very process of information exchange is grounded in the dynamic, unruly and heterogeneous situations of social life, as noted by Cohen. Information privacy is not about principled protections in processes of information exchange. Instead, information privacy regards the boundary management of play and playful spaces for selfhood development.

Cohen's broader notion of information privacy resonates strongly with the collected challenges identified in Chapter 7, particularly regarding the dispersion of boundaries and the absence of control over

20 Cohen, 'Between Truth and Power' 58.
21 Cohen, 'Turning Privacy Inside Out' 13.
22 Cohen, 'What Privacy Is For' 1908.
23 Cohen, 'Turning Privacy Inside Out' 20.
24 Cohen, 'The Surveillance-Innovation Complex' 213.

data generation in complex, socially situated spaces, like the smart home. In the newly developing spaces of the collected world, information privacy as boundary management becomes increasingly important to understand, because the dispersion of traditionally relied-upon boundaries reduces the ability to control information. The notion of informational control in boundary-dispersed environments becomes problematic, giving rise to questions about the control basis of information privacy to preserve spaces for autonomous growth, which Cohen also critiques. An intended outcome of information privacy law thus shifts focus from the protection of individual autonomy to acknowledging the complexity of situated intersubjectivity.

8.3.1   From Autonomy to Situated Intersubjectivity

Situated intersubjectivity is Cohen's conceptual baseplate, and it lies at the heart of everything else that flows from her thinking about privacy. It is used as the primary means to rebut claims founded on liberal ideals that the preservation of individual autonomy is information privacy's goal. Chapter 5's coverage of Stanley Benn's relationship between autonomy and privacy is a case in point. Benn argued that the autonomous human was the embodiment of the rational individual who was freely able to make self-serving choices within the broader value set of society. Information privacy's role is thus to ensure non-interference with the informational decision making of individuals. Information privacy is about the ability to control information and informational access to make unfettered decisions. In this sense, information privacy is inherently about the establishment of the private, in which the claim for a private state signals an intention of privacy. Privacy intrusions therefore emanate in 'unjustifiable interferences into the interests that autonomous individuals have in establishing, sustaining and developing personality and personal relations'.[25]

As noted in Chapter 5, at face value there are crossovers between Cohen's relational characteristics of information privacy and Benn's protection of relational interests. Cohen acknowledges that the ideation of the liberal self is worthy as part of the broader commitment to independence of thought and self-actualisation that run in tandem as a broader societal necessity.[26] Cohen and Benn therefore share some of the same aspirations of selfhood as a foundational element of personal and communal growth. However, while there are similarities

---

[25] Benn, 282.
[26] Cohen, 'What Privacy Is For' 1911.

in aspiration, there are fundamental differences in how those aspirations transpire. Cohen is critical of the liberal tradition of autonomy and the production mode for the autonomous individual, which, she argues, are largely viewed as commonplace in the privacy-related analysis of the legal academy.[27] The dominant model of the autonomous liberal individual is nevertheless important, because it has largely shaped the control basis of information privacy's role as a protector of non-interference. Cohen argues that both the construction of the autonomous liberal self and the role information privacy plays in that construction are problematic.

The first problem arises because the human self does not have an autonomous or pre-cultural core.[28] An autonomous state, unlike what Benn argued, simply does not arise because of non-interference in which reasoned and rational decision making flow because of unfettered freedom. Instead, Cohen has argued that selfhood is a process rather than a state. The process is 'discursive and social', and 'it is informed by a sense of the self as viewed from the perspective of others'.[29] Selfhood and subjectivity are socially constructed through the interaction people have within their networks of relationships, through playful practices and the cultural beliefs they encounter as part of the social experimentation of daily life.[30] As a result, selfhood emerges through active engagement with society rather than non-active and non-interference protections from it. Unlike in the case of the autonomous 'liberal self', selfhood and social shaping for 'the inter-subjectively situated self' are not mutually exclusive.[31] In fact, the opposite is the case. Social shaping takes places everywhere, and always, which means that access to information about one's culture and the surrounding world is crucial to the subjective development of selfhood.[32]

The second problem flows naturally from the first and implicates a certain role for privacy in the role of selfhood development that represents differences between negative and positive dispositions. For the autonomous liberal self, the role of privacy is unsurprisingly defensive and ameliorative.[33] If autonomy flourishes from non-interference, the role of privacy and information privacy is to protect the private spaces in which autonomous growth pervades. The role of privacy, much like the states of

---

[27] Cohen, 'The Networked Self in the Modulated Society' 69–70.
[28] Cohen, 'What Privacy Is For' 1908.
[29] Cohen, 'Turning Privacy Inside Out' 14.
[30] Cohen, 'What Privacy Is For' 1909.
[31] Ibid 1908.
[32] Cohen, 'The Networked Self in the Modulated Society' 70.
[33] Cohen, 'What Privacy Is For' 1908.

solitude or the right to be let alone, as detailed in Chapter 5, is to afford shelter from the pressures of societal and technological change.[34] On the other hand, if selfhood arises from constant social shaping through societal interaction, then the role of privacy, according to Cohen, is markedly different. Privacy cannot be a fixed condition, unlike the negative protector of private space. If an individual's relationship to social and cultural contexts is dynamic, then, as already mentioned, privacy and information privacy must also have the same dynamic characteristics.[35]

Privacy thus shifts to the positive management of physical, spatial and informational boundary strategies that are dynamically flexible in different contexts and social situations. Privacy is not about the negative protection of a given boundary space – that of the private. Instead, it is about the positive management of many boundary spaces, which are always everywhere and emanate constantly, rather than merely at points of liberally oriented, informational decision making. Cohen's role for privacy, as part of subjective selfhood development, is further complicated in data analytic structures because automated logics disrupt the processes of self-formation in unimagined ways,[36] as outlined further on in this discussion.

Cohen's envisioning of the dynamic role of privacy in the context of subjectively situated selfhood speaks closer to the collected challenges outlined previously than does the concept of the autonomous liberal self. Collected spaces, such as the smart home, are sites of such dense sensorisation that the very notion of that space as 'private' is problematic, as it is permeated by the manifest collections of always-on sensor data. As noted in Chapter 7, control-based applications of information privacy become significantly challenged, because the means of gaining control, of building the kind of negatively oriented boundaries envisaged as part of the liberal paragon of privateness, simply do not exist in the smart home. The challenge here also regards the fixed nature of information privacy as control over personal information which is not able to adapt to the complex demands of technological fragmentation.[37] Part of the reason, as detailed later, regards the overt focus on process protections, when those data collection processes have rapidly and markedly changed. Information privacy, as dynamic forms of situated boundary management, is more likely to be able to adapt to these technological changes because they focus less on process regulation.

[34] Ibid.
[35] Ibid.
[36] Cohen, 'Turning Privacy Inside Out' 14.
[37] Bert-Jaap Koops, 'The Trouble with European Data Protection Law' (2014) 4 International Data Privacy Law 250.

The collected challenges and the critique of the control basis of information privacy also gives rise to critical questioning of selfhood development and autonomy protection as an intended outcome. The data collection complexities of the smart home are such that it is questionable whether most individuals could really be deemed to be making free and fully informed decisions about personal information provision based on privacy policies. Again, the complexity of the privacy policies analysed in Chapter 4 is a case in point, especially in relation to Solove's cognitive criticisms of privacy self-management. Moreover, Cohen's assertion that social shaping is constant becomes an important point of consideration. Benn warned of the heterarch dangers of manipulation that could unduly influence free and unfettered individual decision making. These warnings relate to the unseen use of power as malign and manipulative influence over unfettered decision making that would prevent autonomy from flowing.[38] The resultant autonomy protection is a signal of privacy in which the private space of autonomous protection essentially becomes a power vacuum. The vacuum releases objects from the malignantly manipulative forces of powerful others that can exert power over an individual through control over social objects.[39] However, if, as Cohen argues, social shaping is an intrinsic aspect of selfhood, then manipulation is perhaps not the appropriate construct concerning which to consider the role of information privacy and power.

8.3.2   From Power Vacuums to Modulation

The second intended outcome of control-based information privacy law is the amelioration of power imbalances between individuals and data collectors. The control basis of information privacy, as just noted, envisages a primary role of autonomy enhancement through the creation of protective spaces around informational decision-making points, particularly at the point of collection. These points of power vacuum create spaces of non-interference in which a privacy policy, as the primary artefact of understanding, can be neutrally absorbed and processed to fully inform rational and reasoned individual decision making about personal information exchange. Again, Benn's work is instructive regarding the underlying logics at play.

Chapter 5 noted that Benn's state of privacy is one that is private: it is the private space that prevents unwarranted interferences as wilful acts of power intent on manipulation. In turn, those private spaces

---

[38] Zarsky 174.
[39] Benn 266 referring to objects such as 'correspondence, affairs or rooms'.

are dependent upon an individual's ability to control access to private objects. Privacy interests emerge when an individual chooses to embark on a state of privacy – for example, to make some aspect of their life private – or has the power to make some aspect private. Privacy, and its embodiment in the private, is therefore a norm in which rights can be imposed on objects, which signals an expectation of duly-followed behaviour by others. Benn's notion of power can be categorised as 'power over', both in terms of individuals who seek control over objects to make them private and data collectors who seek power over individuals through manipulated decision making that favours their interests. The power vacuum thus allows autonomy to flow by permitting control over objects to make spaces private and to prevent manipulation taking place.

The construct of 'power over' has been a dominant consideration in sociological concepts that seek to understand power in different ways. Forms of 'power over', which form the basis of power amelioration under the control model, are not the only considerations of power relevant to this discussion, as Cohen's concept of modulation outlines. Before we get to modulated forms of power, it is first necessary to consider the multidimensional nature of power and briefly discuss Steven Lukes' seminal work on dimensions of power,[40] to adequately situate the complexity of Cohen's modulated form.

Building off a tradition of political and sociological research that focused on the visible actioning of power by powerful entities, Lukes identified three dimensions of power.

1. The first dimension of power is directly visible and can be observed in the behavioural operation of decision making; for example, A has power over B to the extent that A can get B to do something that B would not otherwise do.[41]
2. The second dimension of power is less obvious and involves the ability of decision makers to make non-decisions which may not be directly observable; for example, A has power over B to exclude B from highlighting issues that would be detrimental to A.[42]
3. The third dimension of power is invisible, and it represents the intrinsic ability of the powerful to shape language, symbols, myths and wants; for example, A has power to shape values and symbols, which influences B to think in a way they would not otherwise do.[43]

---

[40] Steven Lukes, *Power: A Radical View* (2nd edn, Palgrave Macmillan 2005).
[41] Steven Lukes, *Power* (Palgrave Macmillan 1974) 11.
[42] Ibid 20.
[43] Ibid 33.

Lukes argued that the first and second dimensions' focus on actual observable conflict is misleading, as two types of power exist that do not have to involve any conflict: the power to manipulate and the power of authority.[44] Power then does not simply manifest in conflict situations but can be determined through thought control – for example, where A exercises power over B by shaping and determining their needs or interests. This determination of needs is the most 'supreme and insidious' use of power, because it can stop conflict before it arises and does not involve the suppression of any form of surfacing conflict.[45] Actual conflict is not a precursor or a necessary determinant for an exercise of third-dimensional power. Power therefore does not necessarily emerge from decisions alone.[46]

Instead, forms of power can be used to shape the perceptions of individuals so that a grievance may not necessarily surface, as there has been no conception of any alternative. Lukes' third dimension thus focuses on the underlying social patterning which determines the capacity to wield power and highlights systemic forms of power. In the context of networked societies, it has an infrastructural element that is baked into the design of the infrastructure, its operating code and its data collection devices. The third dimension of power, and indeed Cohen's modulation, draws attention to hidden power mechanisms inherent in social relations, of which the social actors are unaware.[47] In other words, Lukes' third dimension of power offers a deeper sociological analysis of the use of symbols in society and the power to shape thought to suppress and control conflict before it arises. Lukes' analysis has been generally conceived as a formative step in the conceptualisation of power. Nevertheless, his work has been criticised, particularly regarding methodological and epistemological considerations.[48] However, the coverage of Lukes' three-dimensional forms of power are introduced here not as a methodology but rather as a means of differentiating the construction of power relevant to control-based notions of information privacy and Cohen's modulation.

Benn was clearly aware of the potential dangers that could arise when powerful forces could manipulate the informational decision-making

---

capacities of individuals for their own ends. Benn's heterarch is a direct clarion call of those dangers. However, as noted earlier, Benn's work is predominantly focused on individual protection against power imbalances along first-dimensional lines. The power vacuum prevents actions of 'power over' by data collectors, which will result in seeming flows of autonomous-thinking from individuals. However, Chapter 5 noted that even though power in the context of information privacy law is treated rather tacitly, it is nevertheless clear that different forms of power-related analysis have already developed. These forms of analysis highlight that the use of power in relation to information privacy law's application can manifest in structural, hegemonic and dispersed forms. As such, the control model's overt focus on power as 'power over' does not pay enough regard to the use of Lukes' 'supreme and insidious' forms of power that are deeply embedded in the infrastructures, logics, social relations and device designs of the collected world. In other words, the control model's focus predominantly regards the first-dimensional use of power, but the real concern of the collected world regards the intertwinement of first- and third-dimensional forms of power, as exemplified in Cohen's modulation.

Modulation offers a more complex frame through which to view the operation of power in the collected world, because Cohen considers power structurally across several different spectrums. These spectrums cover the broader political economy behind ever-cumulative forms of informational capitalism and its concomitant surveillant logics, infrastructural requirements and device designs. In other words, the use of modulated power flows from the macro structures of informational capitalism through to micro activities of sensorised data collections. Power thus flows from the heart of platform metropolises, such as Silicon Valley, directly to the heart of sensorised homes and all other sensor environments of the collected world. However, as Cohen notes, modulated flows of power are different from the types of one-way flow that typify the 'power over' concerns of information privacy's control basis. Rather, modulated power flows in two directions, which is an important consideration in relation to collected consequences.

Modulation is a technique for extracting knowledge insight that is derived from the appropriation of increasing data surpluses in an ever more precise and complete fashion.[49] Its purpose is 'to produce tractable, predictable citizen-consumers whose preferred modes of self-determination play out along predictable and profit-generating trajectories'.[50] 'Modulation'

---

[49] Cohen, 'The Networked Self in the Modulated Society' 74.
[50] Cohen, 'What Privacy Is For' 1917.

essentially refers to the logics of data analytics, which seek to generate ever-increasing forms of acquirable data – mostly in the form of sensor data – and the collection, management and analysis of that data to generate unintuitive insight.[51] The value of Cohen's analysis lies in the acknowledgement that these forms of big data logic and practice generate a set of surveillant data processes which seek to constantly monitor and track individual activities at ever-finer levels. Knowledgeable insight generated from sensor data is then used to continually adjust data generation practices that modify individual behaviour through incessant forms of convenient personalisation, aimed at the provision of more efficient or valuable services.[52]

Similar to the segmentation process of prescriptive analytics, outlined in Chapter 3, modulated forms of data generation offer a Janus-like benefit both to individual data providers and corporate data collectors, like the smart home insurance models of Chapter 4. Individuals receive an apparently seamless service, and data collectors are more able to know more about their customers to better aid service targeting and to better quantify risk based on highly individualised valuations.[53] The power elements of modulation arise because the continuous process of data generation and modification is, at times, undertaken in conjunction with data generated by the data provider, but it is also undertaken 'according to logics that ultimately are outside the subject's control'.[54] First-dimension 'power over' forms are consequently evident because modulation represents a 'mode of governance designed to produce a particular kind of subject'[55] that acts in accordance with data collector needs. However, third-dimension forms of power also emerge because modulation is 'a mode of knowledge production designed to produce a particular way of knowing'[56] that sets the foreground for shaping subjective understandings of individuals. Modulated power therefore has a two-way flow that cements data collection with micro-controlling outcomes through sensorised technology which seeks to achieve social shaping at scale.

Modulation thus 'exemplifies power-as-control, translated into the realm of commercial activity and adapted to the purposes of informational capitalism'.[57] The control element which is redolent in 'power over'

---

51 Andrejevic and Burdon.
52 Cohen, 'Affording Fundamental Rights' 62.
53 Ibid.
54 Cohen, 'What Privacy Is For' 1915.
55 Ibid 1917.
56 Ibid.
57 Cohen, 'The Networked Self in the Modulated Society' 74.

formulations regards forcible individual exclusions from the means
of knowledge production,[58] even though it is predicated on personal
information provided by the individual and, more importantly, exclu-
sion from the means of prediction,[59] which are undertaken outside the
bounds of individual control or cognition. Forcible exclusions emerge
in blunt forms through the industrialisation of manufactured consent[60]
that reduces levels of individual awareness flowing from the embedded
invisibility of sensor collections to the unread, but nevertheless health-
ily fostered, privacy policy. They also emerge in more subtle processes
of continual feedback aimed at further shaping predicted outcomes that
result in prescriptive nudges in line with profit-maximising motives.[61]
Modulation, as a form of 'power over' is consequently 'a highly granular, feedback-driven approach to the study, organization and ongoing
management of populations of consumers'.[62]

Modulation as a mode of knowledge production is more redolent
to the third-dimension applications of power highlighted by Lukes.[63]
It involves the shaping of individual and social consciousness at scale
through global circuits of information flow that serve 'larger constel-
lations of economic and political power'.[64] These patterns further
disperse data collection capabilities, which in turn promote private
sector power over information at the expense of states.[65] The increasing growth of private economic power thus leads to new forms of
informational domination in the form individualised predictions and
prescriptions that are 'the natural and normal outcomes of market processes'.[66] In other words, the predicted outcome and the prescriptive
nudge are the normalised and internalised core aspect of the collected
world. This new form of modulated normalisation also requires 'a distinct ideology and appropriates for its practitioners a particular kind
of power over knowledge'.[67] At the heart of modulation lies big data
logics that equate 'information with truth and pattern-identification
with understanding'.[68]

[58] Cohen, 'The Surveillance-Innovation Complex' 64, 66.
[59] Cohen, 'The Networked Self in the Modulated Society' 71.
[60] Cohen, 'What Privacy Is For' 1917.
[61] Ibid.
[62] Cohen, 'The Surveillance-Innovation Complex' 64.
[63] Note that Cohen states that modulated power is not just of the hegemonic variety.
Cohen, 'The Biopolitical Public Domain' 13.
[64] Cohen, 'Affording Fundamental Rights' 58.
[65] Ibid 64.
[66] Ibid 66.
[67] Cohen, 'The Networked Self in the Modulated Society' 74.
[68] Cohen, 'Between Truth and Power' 65.

The two-way flow of modulation becomes important at this point. The modulated form of norm establishment and the processes of data collection and analysis become the means of shaping citizens into citizen-consumers. Selfhood is no longer a healthy requisite for individual and societal flourishing. Instead, it is a modulated mode of data-based shaping in which selfhood is determined 'along predictable and profit-generating trajectories'.[69] A corollary shaping effect emerges regarding the use of data-collecting technologies. As noted throughout the book, it is the sensor-based devices that enable sense-making in the collected world. It is these devices that we use to increasingly organise our worlds. However, the same devices are used to subtly shape the way we make sense of the world and the way that data collectors make sense of us; hence the real power of modulation's two-way flow, because 'over time we come to perceive the world through the lenses that our artifacts create'.[70]

Under modulation, this is where the ultimate source of power lies – beyond 'power over'. It is a third-dimensional form of power that seeks to adhere ubiquitous collections of sensor data, and the continuous forms of tracking that follow, as an entrenched and irreversible norm of the collected world. Information privacy protections as ameliorations of power do not regard the generation of power vacuums at points of informational decision making. Instead, modulated forms of power emerge through their 'plasticity' and offer multiple points of infrastructural entry for private power to unfold.[71] These multiple points of entry relate more closely the complex data anatomies of smart environments detailed in Chapter 3's analysis of the smart home. Dense forms of sensorisation are the multiple points of entry that facilitate the generation of data from which modulated personalisation strategies perpetuate. Sensor entry points and the power flows that emanate are thus conjoined. All of this makes the type of power to make objects private, as envisaged by Benn, problematic.

The many communication networks of the smart home provide the infrastructural frame, but it is the sensor that is the initiating power component. Without the sensor, the frame cannot function effectively and produce the individualised predictions and prescriptions that fuel modulated profit. As Cohen notes, these systems can only work if they are sufficiently efficient at returning commercial gain.[72] In other words, the processes of

---

[69] Ibid 65–6.
[70] Cohen, 'What Privacy Is For' 1913.
[71] Cohen, 'The Networked Self in the Modulated Society' 77.
[72] Cohen, 'What Privacy Is For' 1913.

data analytics are forever required to make the unintuitive insights that its beholden logics necessitate, including ever more sophisticated forms of sensor data collection, ever-finer groupings of segmentation and ever more subtle prescriptions. In effect, the colossal infrastructures of the collected world are dependent on the proliferation of tiny sensors to function.

These implications are considerable when the fragmented structures of collected places such as the smart home are obviated. Chapter 4's coverage of the Platform Entity model is a case in point. The model will grow in prominence from the increasing convenience drift that is powering consumer use of smart home voice assistants, such as Amazon Echo and Google Home. These devices are fast becoming the dominant form of smart home controller. If that is the case, then the unplanned protections that are currently available through fragmentation will reduce as smart home data is increasingly collected, recorded and analysed through one medium. In effect, the titan platforms will be collecting data from millions of home-based sensors worldwide. At that point, the path to modulation moves from conceptually concerning to real, because the most powerful data-collecting organisations will be able to set the conditions for smart home knowledge production and the mode of predicted and prescripted governance.

Cohen's modulation is therefore important to the overall critique of the control basis of information privacy law's intent of power amelioration. It points towards the role of the sensor as a multiple and contingent entry point of power that is beyond the scope of any power vacuum based on the ability to make an object or space private. If that is the case, then information privacy law must look more carefully at how sensors operate, collect data and shape our understandings of the collected world. Thus far, as Cohen argues, traditional constructs of liberalism have typically not paid attention 'to the processes by which power relations are encoded in technologies and artifacts',[73] such as sensors in the data-collecting devices of the collected world. These considerations point towards a greater role for information privacy law at the point of collection and different modes of application that look beyond the constraints of transactional operation founded on process-based use of principled protections.

8.3.3   From Transactional Operation to Boundary Management

As previously mentioned, the control basis of information privacy results in modes of transactional operation based on tradable exchanges of personal information. The primary source of information privacy

73 Cohen, 'The Networked Self in the Modulated Society' 76.

protection – the principled approach of life-cycle regulation, such as the US FIPPs or the Australian APPs – regulates the personal information exchange processes and legitimate information privacy's transactional modes of operation. Information privacy law obligations are thus process based. Protections seeks to imbue fairness as a normalised component of transactional operation by ensuring that individual data providers retain degrees of control over personal information. Individual control and process fairness are enmeshed together and form the basis of legal and conceptual responses that tacitly enhance autonomy and attempt to ameliorate the power of data collectors. However, Chapter 7 detailed the challenges for the control basis of information privacy that arise from the smart home's fragmented structures and its dispersion of boundaries that greatly reduce individual opportunities over process control. In fact, all parties, to a certain degree, currently lose control over transactional operations, particularly around data collection.

Cohen's requirement of information privacy as a regulatory mode to interrupt modulated practices is fundamentally different to the process-oriented modes of transactional operation that pervade in the control model. Information privacy is not a static good to be traded off against other possible goods, such as the provision of service convenience. Instead, it is an 'environmental condition and a related entitlement (or set of entitlements) relating to that condition'.[74] It is also important at this point to remind ourselves of the basis of Cohen's definition of privacy. It regards boundary management that creates spaces of play in which persons undertake activities of self-definition and understanding. Privacy is about breathing space for play as selfhood development. Information privacy requires 'affirmative measures designed to preserve and widen interstitial spaces within information processing practices'.[75] Information privacy law protections against modulated forms of power are about protecting the spaces between the demarcated boundaries of transactional process rather than the process points themselves. Hence Cohen's focus for information privacy is environmental instead of processual. It regards boundary management of gaps and spaces rather than the protection of process points. It is these gaps and spaces that enable selfhood development through play with life situations. And it is the gaps that prevent the seamless harms of modulation to emerge.

More precise modulation of information flows, aided by better-informed input from consumers, will not provide more privacy or better privacy. If privacy regulation is to provide effective shelter for

---

[74] Cohen, 'Turning Privacy Inside Out' 20.
[75] Cohen, 'The Networked Self in the Modulated Society' 77.

socially situated processes of boundary management, we will need to acknowledge that privacy is the opposite of modulation and can exist only to the extent that processes of modulation are gap-ridden, transparent and incomplete.[76]

If Cohen is indeed right, that information privacy is about boundary management of socially situated processes, then information privacy law must also aim to operate in the gaps as well as at the process points. Operation, in this sense, is not simply about creating and enforcing boundaries as a protection which is the basis of the control focus of information privacy, such as Benn's privacy powers to make an object private. Nor is it about the control basis of life-cycle protection that creates principled boundaries of application and pathways of individual participation and legal obligation between those boundaries. Instead, it is about the explicit acknowledgement that information privacy is relational. Therefore protections, in the form of boundary management, are needed for the interstitial spaces between structures, objects and processes,[77] because that is where subjective selfhood forms and emerges. As such, the nature and quality of the interstices are of paramount concern because they will assist in shaping individual selfhood.[78] The role of information privacy law is therefore to interrupt the seamless two-way flows of modulation in sensing infrastructures dependent upon always-on sensor data collections from the collected world.[79]

However, as Cohen rightly argues, the current regulatory target of information privacy law is not the interstitial spaces between process points; it is the process points themselves. Even the GDPR, with its added rights-based foundation, nevertheless solely regards the processing of personal data. Anything that is outside the processing activity or the confines of personal data will not give rise to coverage. Modulated harms, under control-based information privacy law, thus timidly arise as the traditional evils of unauthorised disclosures or secondary uses that disregard the power-related elements of modulation. Future modulated harms will continue to arise in ways different to those envisaged by information privacy law.

Some harms will emanate in processes of transactional operation, but others will emerge in the newly formed spaces and gaps of selfhood that sit between process points and beyond. These harms are emerging already as problems with re-identification possibilities from

---

[76] Cohen, 'What Privacy Is For' 1930.
[77] Ibid 1931.
[78] Cohen, 'Turning Privacy Inside Out' 21.
[79] Cohen, 'The Surveillance-Innovation Complex' 219.

de-identified data,[80] pattern recognition that reveals sensitive aspects of a person's life without directly collecting data from them,[81] inferred predictions[82] or the prescriptive nudges that seek to normalise these modulated practices as everyday life.[83] Information privacy, removed from its shackles of process control, will need to develop new regulatory responses that detach privacy from individual control mechanisms and reorient it towards the environmental conditions of socially reflective life.[84]

Cohen argues that this can be achieved through new strategies based on semantic discontinuity that directly attend to 'the essential roles of gaps, barriers, breakdowns, and failures of translation in producing the conditions that render selves incomputable'.[85] Instead of information privacy law's process of principled protections that provide legal obligations to balance rather than prohibit seamless exchanges of personal information, strategies of semantic discontinuity do the opposite, as they promote gap production as part of increasingly automated structures of data collection and exchange. Semantic discontinuity is therefore 'the opposite of seamlessness',[86] because it requires respectful and careful consideration of the role of gaps and spaces in data exchanges.

Modulated power arises from the ability to crush process spaces so that the interstitial boundaries of time and space no longer exist. The crushing of process space and the concomitant gaps around it means that there are few, if any, boundaries around interstitial spaces of selfhood development. Collection is analysis. Analysis is outcome. Prediction is prescription. Sensor is the nudge. In other words, there is only one continuous process of automated data exchange, which makes it easier for the two-way flow of modulated forms of knowledge production and prescriptive governance to foster unchecked.

Semantic discontinuity creates new forms of check to prevent the crushing of process space through the promotion of seamful stopgaps.[87] As such, it is 'a right to prevent precisely targeted individualization and continuous modulation'.[88] Barriers, supported by new forms

---

[80] Paul Ohm, 'Broken Promises of Privacy: Responding to the Surprising Failure of Anonymization' 57 UCLA Law Review 1701.
[81] Zuiderveen Borgesius.
[82] Solon Barocas and Andrew Selbst, 'Big Data's Disparate Impact' (2016) 104 California Law Review 671.
[83] Yeung, '"Hypernudge"'.
[84] Cohen, 'Turning Privacy Inside Out' 3.
[85] Ibid 21.
[86] Cohen, 'What Privacy Is For' 1932.
[87] Cohen, 'Turning Privacy Inside Out' 24.
[88] Cohen, 'The Networked Self in the Modulated Society' 78.

of information privacy law, are required to realise the right to not be continually modulated. These barriers are 'fixed, definite, cumulative and redundant' to information processing and therefore create the necessary spaces for breathing room and play to flourish as part of selfhood development.[89]

Semantic discontinuity strategies aid personal boundary management of space that is beyond the transactional operation of principled protections. Cohen contends that they are counterintuitive, because they work against the efficiency basis of the transactional approach which is inherently predicated on the balancing mechanisms of information privacy between individual provider and data collector. These balancing mechanisms intentionally require efficient processes of exchange to facilitate data collector needs. However, efficient transactions, in the form of the increasingly automated processes of data collection that make up the collected world, take efficiency to another level based on forms of hyper-individualised service provision based in turn on continuous collections of personal information.[90]

Semantic discontinuity's boundary management aims are deliberately inefficient in focus. Gaps and spaces derived from its boundaries seek to diminish the efficiency of the seamless, two-way flows of the data exchange process of modulation. Different boundary management strategies thus arise – for example, record-keeping gaps that make automated assemblages of complete and continuous collections of personal information more difficult to realise.[91] This means new information privacy rights of incomplete processing that respect a right to be remembered imperfectly or imprecisely.[92] All of this leads to an overarching protective sphere that obviates predicted and prescriptive judgements. Semantic discontinuity, at its heart, therefore regards a right to not be judged based on the recorded activities of everyday life, which assists in defining requisite 'information rights and obligations and establish[ing] protocols for information collection, storage, processing, and exchange'.[93]

Cohen's semantic discontinuity heralds a move from subject-centred control of discrete information flows to condition-centred frameworks[94] that consider more heavily the role and design of technology as part of

---

[89] Ibid.
[90] Yeung, '"Hypernudge"' 122.
[91] Cohen, 'The Networked Self in the Modulated Society' 77.
[92] Ibid.
[93] Cohen, *Configuring the Networked Self* 239.
[94] Cohen, 'Turning Privacy Inside Out' 24.

the processes of social shaping and modulation. The detachment of information privacy from processes, including process actors, means that information privacy law can reorient its focus 'toward environmental considerations' that open greater opportunities of operational possibility.[95] Those new areas of operational possibility regard the establishment of gaps in legal, technical and informational architectures that 'create the breathing room within which the play of everyday practice occurs'.[96]

A problem arises, then, from current information privacy practice, because the gaps operate the other way around. Chapters 4 to 6 highlighted the problematic application of privacy policies and the gaps of understanding they generate either by their intentional vagueness or by their inability to adequately capture and relate the true complexities of data exchange mechanisms. In other words, the privacy policy, particularly in relation to sensor data collections, as noted in Chapter 4, creates gaps of unawareness which are the very opposite of what Cohen's semantic disclosure requires. These awareness gaps could be classified as modulation enhancing rather than preventing.

Gaps and spaces created based on semantic discontinuity are process-obscure spaces, because they reside at the edges of process. They are definable but deliberately confused spaces in which play can arise. Again, in many ways, Cohen's role for information privacy law, which has a greater cognition of environmental conditions, relates more accurately with the smart home considerations outlined in the first part of the book. The fragmented nature of the smart home naturally creates gaps and spaces, but the boundaries around them are ill-defined. As such, they are boundary-less rather than boundary-free spaces.

The discussion on semantic discontinuity and modulation has shown that the gaps of the collected world are both a protection and a threat. The gaps protect against the development of a seamless modulated form. However, the threat arises from the nature of sensor data collections, which are better able to define the spaces of protective gaps that can then be owned or filled by modulated forms of transactional operation. Again, Chapter 4's discussion of the Platform Entity model is a case in point, where the extensive implementation of Google Home or Amazon Echo devices gives rise to the increasing scope of condition setting by the most powerful platform entities. The two-way flow of modulation through sensor data collections therefore poses the threat that gaps can be identified, as part of continuous knowledge production

---

[95] Ibid 21.
[96] Cohen, 'The Networked Self in the Modulated Society' 78.

of always-on collection processes, and that they can then be conditioned as part of prescriptive governance of individual agency.

The smart home insurance models are another case in point, where the requirement for preventative measures, as a way of securing better individual premiums, means that certain activities, rather than the individuals themselves, are defined as valuable by predictive models. These range from the obvious actions of keeping the home adequately secure to more obscure micro activities such as ensuring that battery life in smoke detectors is maintained. Again, the fact that a sensorised smoke alarm can collect data about battery life patterns means that more micro activities can be detected and monitored for prescriptive outcomes. The cycle of smart home data collection, prediction and prescription is thus circular and continuous and covers an ever-increasing range of activities and behaviours due to the implementation of a greater number of sensorised devices and infrastructures.

Accordingly, forms of semantic discontinuity should not be solely about computability, as a form of data analysis. Instead, it should also arise from actions that involve collectability, because computability can only arise out of data collection. The sensor, the data it produces, and the ultimate analytical outcome are insuperable. Data collection and its role in social shaping through modulation therefore needs much greater recognition under information privacy law than currently exists. The overt focus on disclosure mechanisms as a means of ameliorating power imbalances thus misses the important role that information privacy law should explicitly have in relation to social shaping.

8.3.4   *From Information Asymmetries to Social Shaping*

Different parts of the book have highlighted the important role that privacy policies play in control-based models of information privacy. The privacy policy has a cornerstone position that provides a foundation for further protective developments involving the provision of informational spaces for autonomous decision making and to rebalance information asymmetries. It also plays a significant role as a primary point of rationality generation in the control model's transactional modes of operation. The privacy policy, in its role as an integral information disclosure mechanism, seeks to ameliorate information asymmetries and thus also plays a major part in the establishment of power vacuums. It therefore has a two-fold requirement: to generate individual understanding to fully inform decision making about personal information exchange and to place obligations on data collectors to provide meaningful information. Those obligations form a frame for reducing asymmetries by requiring data collectors to provide specified types of information, such as the purpose of collection and subsequent uses. The latter requirement also has a key role in the formulation of the power vacuum, as designated obligations attempt to remove the potential power advantages that data-collecting entities have over individuals.

However, Cohen's work shows that the power-related issues of modulation are broader than the scope of the power vacuum for autonomous decision making. Consequently, the control model's intended outcome of ameliorating information asymmetries also must be re-evaluated. Like the power-related analysis relevant to forms of modulation, information privacy law should be reconsidered to have a more prominent role regarding the character of social shaping, as highlighted by Cohen. This new implementation outcome regards a range of environmentally constituted and design-focused measures that connect the material conduct of devices directly to information privacy law's regulatory ambit. It is essential to do this, because data collecting devices, and the sensors they behold, can shape individual and collective understanding of the social and physical world. A greater focus on the material effects afforded by devices is therefore a reflection of the foregoing discussion relating to the move from process regulation to boundary management of gaps, which requires a greater emphasis on relational and environmental considerations of the collected world.

As already noted, Cohen regards the affordance nature of privacy as an 'environmental condition and a related entitlement (or set of entitlements) relating to that condition'.[97] Information privacy applications, formulated in terms of material effect, focus on technological applications that are informed by the context of the 'built environment's systemic tolerances and prohibitions'.[98] Information privacy law protections should derive from the 'constraints and affordances of the physical environment' as well as institutional legal structures, because that is where prime forms of social shaping take place.[99]

In other words, we understand our role in placating and preserving information privacy law protections, not from exogenous forms of law, but from endogenous interaction with technology and devices in socially situated settings. To return briefly to Benn and privacy powers over objects, Cohen contends that the two-way flow of modulation operates directly between individuals and devices. As such, the individual does not simply retain agency powers of privacy over objects,

---

[97] Cohen, 'Turning Privacy Inside Out' 20.
[98] Ibid 19.
[99] Cohen, 'Affording Fundamental Rights' 85.

such as devices, to make them private; rather, power flows the other way around. The device has the shaping power to determine individual understandings of when something could be private and can therefore be configured in a way that does not emanate in a claim of privacy.

It is already clear that these considerations are different to the structures of notice and consent that prevail in the control model of information privacy. Cohen differentiates clearly between rights to privacy, as a liberty-based formulation, and information privacy rights, which are entitlements 'better suited to articulation within an affordance-based discourse'.[100] Information privacy law rights, as affordances, require a shift in focus to consider in much greater depth infrastructural and operational details[101] – such as the gaps and spaces that warrant care and which exist at the edges, or outside, of data exchange processes. The application of information privacy is therefore 'first and foremost a matter of design'.[102]

The position adopted in this book places a different emphasis than that of Cohen about the relationship between privacy and information privacy law. As noted already, it is argued here that the separation of one from the other is not clear-cut and, more importantly, is not desirable. The intended outcomes of information privacy law highlight the cross-over between the four conceptual themes detailed in Chapter 5. These, in turn, cross over with broader notions of privacy, such as personhood. Moreover, if Cohen is right in the assertion that social shaping takes place everywhere and that the shaping involves a modulated power characteristic that forms of privacy law, including information privacy, need to be aware of, then it becomes problematic to ascribe a role for information privacy as solely affordance based and one that is therefore consigned only to design considerations. A greater consideration about the role of law in allowing enhanced forms of boundary management to preserve selfhood warrants a reconfiguration of information privacy and its relational application between individuals, devices and infrastructures. However, that does not mean it has nothing to offer in terms of extending or critiquing legal and political debates about the role of privacy as a liberty-based formulation.

That said, however, Cohen's critique of the problematic use of consent is a strong one. The role of consent 'as a legitimating condition for satisfaction of data protection obligations' is problematic not solely because it 'is a liberty-based construct'[103] but because it is not designed for the

---

[100] Cohen, 'Turning Privacy Inside Out' 19.
[101] Cohen, 'Affording Fundamental Rights' 86.
[102] Cohen, 'Turning Privacy Inside Out' 21.
[103] Ibid.

greater scope of application required to interrupt modulated forms of power which aim to socially shape individuals through sensors, devices and infrastructures. In the collected world, networked artefacts and technologies have the capacity to configure their users, as they encourage individuals to behave in certain ways over others.[104] The most prominent are design features that by default are intended to generate, collect and exchange data.[105] As noted earlier, the use of consent in this regard pertains to binary, one-off decisions, but the consequences of those decisions are forms of continuous, pervasive and changeable tracking and analysis. It is in the world of post-consent provision that the true heuristic capabilities of device shaping unfolds, in which the world is viewed increasingly through a lens afforded by technology, to the extent that we eventually take this provided outlook for granted.[106]

Issues of social shaping have thus far been considered at a micro level, involving the limits of consent provision by individuals and the role of broader environmental legal considerations that are more relational in nature. Cohen's work on the biopolitical sphere also considers issues of social shaping and modulation at the macro level, where forces of private power shape understandings of public towards the exigencies of political economy based on the bioprospecting of personal information.[107] The biopolitical domain becomes the repository of raw data material that fuels the two-way flow of modulation, at scale. Cohen argues that the domain's construct shapes understanding about the role of personal information necessary to fulfil the essentially private activities of commercial production.[108] It is therefore a power act and defines modulated activities that flow from it.

The biopolitical domain's power act regards the allocation of informational objects as a public resource. It is the inverse power act of Benn achieved at massive scale. Benn argued that privacy powers emanate in the ability to make objects private. The biopolitical domain achieves the opposite. It makes all informational objects public, thus denying a countervailing individual privacy power based on the construct of private. Consequently, personal information is constituted in two ways in the biopolitical sphere. First, personal data is available for resource allocation that potentially generates production value.[109] It could potentially be

---

104 Cohen, 'The Networked Self in the Modulated Society' 68.
105 Mireille Hildebrandt, 'Law as Information in the Era of Data-Driven Agency' (2016) 79 The Modern Law Review 1.
106 Cohen, 'The Networked Self in the Modulated Society' 69.
107 Cohen, 'The Biopolitical Public Domain' 213.
108 Ibid 214.
109 Ibid.

argued that the control model's transactional mode of operation accelerates such notions, given the constructions of personal information as a tradable commodity, even in purer exchange senses. Second, personal information is raw and can be harvested for a broader means of informational production.[110] Personal information is therefore an unowned resource that can be appropriated at will either as 'an asset in itself of as an input into profit-making activity'.[111]

An important aspect of the biopolitical domain is scale, as personal information is extracted and exploited at industrialised levels through 'information age refineries'.[112] The industrialisation of informational production means that the role of a privacy policy as part of an information disclosure mechanism to ameliorate asymmetries becomes subsumed into the broader profit maximisation ecology. At one level, the privacy policy remains part of transactional individual data exchange envisaged by information privacy law's control basis. However, at a deeper level, that of the industrialised extraction domain, modes of personal information exchange become about something else – namely, the continual reconstruction of an unbounded informational market that seeks to redefine every individual data exchange into a mode of expression and understanding of what that trade is. Accordingly, in Cohen's biopolitical domain, the shaping of one data exchange is the shaping of all, achieved by the mass production of data doubles that shape, define and prescript data replicas of individuals into data castes of segmented populations.[113] Human behaviour and preferences, through data-double analysis, become 'calculable, predictable, and profitable in aggregate' to enable trade in populations.[114]

The scale of the biopolitical domain is key in relation to aspects of modulated knowledge production that result in power-related intentions of social shaping.

The idea of a public domain of personal information alters the legal status of the inputs to and outputs of personal data processing. In that sense, it is relational and distributive: it both suggests and legitimates a pattern of appropriation by some, with economic and political consequences for others.[115]

Chapter 4's coverage of the US smart home privacy policies could be read as a pattern of appropriation that attempts to alter the legal

110 Ibid.
111 Ibid 215.
112 Ibid 226.
113 Ibid.
114 Ibid.
115 Ibid 230.

status of personal information in its limited designation of protected information as PII. Those policies indicate that sensor data from the relevant devices is not PII and therefore can be traded in distributive networks that seek to accumulate mutual benefit from the selected data of individuals. The selected data, to use Cohen's framing, is the data double that is then refined for further purposing. The refinement is legitimated by individual consent, predicated on privacy policy understandings and based on the shaping of sensor data as non-PII. The biopolitical considerations that arise consequently do not regard information asymmetries, but instead relate to social shaping at an industrialised scale that includes the power to shape legal understandings of the basis and function of information privacy protections. One of the strengths, therefore, of Cohen's work is that it can be used to expose the embedded logics of the biopolitic and modulation, which, again, has implications for the intended outcomes of information privacy law's control basis.

8.3.5   From Balancing Mechanisms to Exposing Modulation

The final intended outcome is information privacy law's inherent balancing mechanisms. As noted already, the control basis of information privacy provides mechanisms to balance and preserve the interests of individuals, collectives and societies. The balancing of interests is predominantly based on notions of fairness, legality and, in the EU, proportionality.[116] The foregoing coverage indicates that the power-related consequences of Cohen's modulated form of informational capitalism are such that it is questionable whether interests can or should be balanced equally as an intended outcome of information privacy law. Put simply, the newly developing ability of corporate data collectors to conduct social shaping at scale, as a basis for modulated forms of power, requires information privacy law to expose the underlying logics of informational capitalism, as opposed to balancing unequal interests. Information privacy law's role in this regard is to raise individual and societal levels of awareness about the industrialised, infrastructural context of biopolitical forms of modulated power.

Information privacy law's role in exposing modulation would highlight the varied effects of modulated practice, as covered previously. Thus, for example, it would assist in highlighting Cohen's important insight that networked artefacts and technologies configure their users

---

[116] Bygrave, *Data Privacy Law* 146.

and that this configuration mediates 'the ways in which we make sense of the world, organizing our perceptions and supplying heuristics'.[117] As highlighted earlier, Lukes' third dimension of power becomes increasingly more relevant here, because continuous analysis of constant sensor data collections becomes the dominant lens for us to make sense of ourselves, our devices and our worlds. At that point, sensor collections, predicted analysis, prescriptive outcomes and the 'settled habits' of individuals[118] all become a seamless flow that in turn becomes increasingly inseparable from one another by time, by process or by space. Device sensorisation of the home therefore becomes the taken-for-granted form of monitoring, tracking and shaping that could be the new reality of everyday life in the smart home.

Information privacy law based on new principles such as semantic discontinuity and operational accountability, as detailed in Chapter 9, could play a significant role in exposing the embedded and invisible practices behind modulation. However, this new exposure cannot be supported through current forms of mechanistic protection of process such as the privacy policy. The deeply embedded nature of sensor collection requirements are such that legally compliant data collection explanations of privacy policies are not an effective awareness-raising mechanism. Even if awareness raising is achieved, the privacy policy points the cognitive consideration of individuals to an outmoded process of data collection, to fulfil the increasingly redundant outcomes of autonomy enhancement and the amelioration of information asymmetries. In other words, the privacy policy is the wrong tool for the wrong job when information privacy law must face up to the modulated consequences of the collected world. Instead, as detailed in Chapter 9, information privacy's role should be in exposing the logics and practices of modulation regarding design features, based on semantic discontinuity and operational accountability, which extend legal protections from processes to the vital gaps and spaces where playful selfhood emerges.

However, the playful self under the logics of modulation is a data-double asset rather than a subjectively developing person. Chapter 7 already noted that the distinction between the individual and the data derived from them is being continuously narrowed in the smart home. Cohen's analysis of data doubles is important in this regard because it highlights the heavily industrialised nature of the data tranche, or caste, as is used previously, which formulates the basis for probabilistically determined purchasing and risk profiling at a personalised and at a population level. Doing so exposes

---

[117] Cohen, 'The Networked Self in the Modulated Society' 68.
[118] Ibid 69.

some of the true powers of modulated practice and its seamless ability to analyse the continuous behavioural flow from the one to the many. At the same time, prescriptive outcomes are algorithmically fed back to the 'ones' as a process of individual, tranche and societal shaping.

The probabilised data double becomes the asset to trade. In that sense, the new oil is indeed personal information, but its extraction requires drilling into the bore holes of sensor data collection to find the rich elixir of activities, behaviours and emotions that represent the social complexity of individual, family and communal life. Cohen's assertion of data extraction and refinement is an important exposure heuristic that highlights that sensor data collections such as those of the smart home do not just entail collections of data in personal information exchange processes. Rather, as noted in Chapter 7, valuable insight is not just about the collection and analysis of sensor data per se; it's about the ability to formulate an aggregated placement of a data double in a broader predicted population, as demonstrated by the data exchange partnerships of both the Partnered Intermediary and Platform Entity models in Chapter 4. The 'limited data' for exchange formulates the data double, and the 'premium offer' becomes the broader predicted population. However, information privacy law based on transactional operation will only reinforce these modulated tendencies rather than expose the deeper, underlying logics of modulation.

Exposure, in this sense, is in its most painful yet rewarding form. Self-reflective critique of information privacy law's own restrictions is necessary if it is to have a stronger exposure role of modulated practices and logics. That is a significant task, because many of the perceptual bases of information privacy law are considerably challenged by the consequences of the collected world. If Cohen's analysis, and thus the arguments put forward in this chapter, is reflective of the actual challenges that will arise from the collected world, then at some point soon, decisions will have to be made as to whether to reformulate the precepts of information privacy's predominant control model or abandon it altogether. These concerns, of course, are not standalone to this book. Instead, they echo many previous voicings of this exact issue. However, if modulated power is indeed the principal challenge, then information privacy law's role in exposing that power will have to take greater prominence.

A greater exposure role of power may not be as daunting as other, future forms of information privacy law reform, such as the move from process protection to 'affirmative measures designed to preserve and widen interstitial spaces within information processing practices'.[119]

---

[119] Ibid 77.

Cohen's principle of operational accountability, particularly in the form of transparency, highlights that current information privacy law has exposure characteristics, but it seeks to wield these with inadequate tools. For example, enhancing individual access to personal information,[120] such as through APP 12, provides insight into collector handling practices, but it is not enough to expose the underlying logics of modulation. Exposure, in this sense, is a matter of institutional and device design rather than one of informational access or disclosure.

[T]he principle of operational transparency suggests that both regulators and designers of networked digital artifacts and interfaces should experiment with ways of disrupting the comfort zones produced by processes of modulation, drawing attention to their existence, and providing individual citizens with the resources to interrogate modulation's logics and effect.[121]

The exposure capabilities of future information privacy law reforms will therefore be endogenous in nature, as they will focus on institutional values about the design affordances of sensorised devices. As noted previously, understanding the role of the sensor is vital in exposing the two-way flow of modulation that is dependent upon sensor-generated data for knowledge production and the sensorised device for prescriptive delivery. Future forms of information privacy law will therefore need to incentivise institutional tinkering[122] as a form of operational transparency and as an ongoing process of semantic discontinuity to ensure that design affordances indeed have the effect of interrupting modulation flows. Exposure of modulation has a strong institutional component that is ultimately representative of the institutional values of the data collector. As discussed in Chapter 10, the regulatory basis of information privacy law shifts from institutional compliance with information exchange processes to institutional articulation and exposure of privacy values. In that sense, institutional tinkering plays a vital role in value development that flows through data collection designs and interrupts modulation by exposing its practices and logics.

8.4   Conclusion

Chapter 8 has undertaken two principal tasks: to critique the control basis of information privacy law and to provide a potential path of reform based on Julie Cohen's sophisticated analysis of the power implications

---

[120] Ibid.
[121] Ibid.
[122] Cohen, 'What Privacy Is For' 1918.

of networked societies. The five intended outcomes of control-based information privacy law are significantly challenged by the consequences of the collected world, as outlined in Chapter 7. Cohen's work is then used to further critique the control model and to demonstrate a future role of information privacy law as an interruptive response to modulated forms of power. In doing so, the five intended outcomes of control-based information privacy move markedly to form a conceptual response against modulation that covers a range of different reforms. Underpinning this conceptual movement is the shift from a subject-centred approach to a condition-centred approach that places much greater protective emphasis on the gaps in and around processes of personal information exchange, as opposed to the process itself. Cohen's analysis prompts positive and careful consideration of those gaps, because these are the spaces where selfhood develops. They are also the spaces where modulated concerns and semantic discontinuity solutions will arise. All of this would ensure that the seamless, uninterrupted flows of data collection and analysis that produce the conditions for modulation are exposed as the primary actions of 'power over' and hegemonic construction that formulate the political economic base of informational capitalism.

The final substantive chapter of the book, Chapter 9, now outlines some potential reforms for current applications of information privacy law. Before we get there, it is important to note that information privacy law is not simply going to disappear overnight or any time soon. The advent of the GDPR and the law's global spread will ensure that it continues to survive into the future. The reforms suggested, therefore, seek to reformulate two crucial aspects – namely, personal information and robust collection principles. The reformulation follows the conceptual analysis of Cohen's work in this chapter and seeks to apply the principles of semantic discontinuity and operational accountability as means of interrupting modulated practices and logics. The result would be new forms of information privacy that are explicitly designed for the power-related consequences of the collected world.

9 Using Information Privacy Law to Interrupt Modulation

9.1 Introduction

Chapter 8 outlined a reformulated conceptual basis of information privacy based on Cohen's sophisticated analysis of modulated power in the types of network societies that make up the collected world. These conceptual considerations would have a significant impact upon the application of information privacy law principally based on mechanisms of individual control in personal information exchange processes. Bringing the embedded consequences of modulated power to a visible surface would require changes in both the construction and application of information privacy law for it to have a more explicit power-related focus. This chapter therefore considers some tentative reforms of information privacy law to facilitate a possible transition from its control base of principled protection of process to one that directly involves structured interruptions of modulation.

As noted in Chapter 8, the control basis of information privacy law is not going to disappear overnight. Its mode of regulatory intention will remain the dominant form of information privacy law for the foreseeable future. Information privacy law reforms should thus seek to reformulate or augment existing legal structures. Chapter 8's coverage of the trials that the five intended outcomes of information privacy law will face against collected challenges, and the conceptual considerations that flow from Cohen's work, reveals a long road to travel from control over personal information to interruptions of modulation. However, while such a journey is likely to be long, and probably winding, there are several reform measures that could be undertaken to start that evolutionary change. It should also be noted that Cohen has so far not clearly articulated specific legal proposals.[1]

---

[1] Cohen, 'The Networked Self in the Modulated Society' 77.

To provide a starting point on the journey, this chapter puts forward legal design points for future information privacy law reform that encapsulate the differences between the control basis of information privacy and its transition towards an interrupter of modulation. The four points represent some base precepts that would be required to move information privacy law towards a more explicit power-related focus. They would be prime components of reformulated information privacy law that could guide legal and regulatory responses to support the move from autonomy to situated intersubjectivity, power vacuums to modulation, transactional operation to boundary management, information asymmetries to social shaping and balancing mechanisms to exposing modulation. The design points would bridge conceptual reformulation and the legislative or regulatory responses required to action a new conceptual form.

The relevant coverage that follows leads with the section heading of 'some' rather than 'four' regarding the identification of key design points. As with most initial attempts at conceptual reformulation of law, it is highly likely that the consideration put forward here is incomplete. As such, it is likely that further design points could be identified. However, it is posited that the four points cover the top-level requirements of change, as indicated in Chapters 7 and 8. These are the baseplate necessities, but it is acknowledged, and hoped for, that existing design points will be refined and that further points will be identified as flowing from the book's overall analysis. With that caveat in mind, the four design points identified thus far are (1) a focus on regulating interstitial gaps as well as process points, (2) information privacy law protections based on seamful interruptions of modulation, (3) the expansion of boundary-building options and (4) exposure mechanisms to surface modulated practices and logics.

Four information privacy law reforms are then put forward based on the application of the design points and consideration of Cohen's principles of semantic discontinuity and operational accountability. The first reform has a broader societal consideration regarding the role of information privacy law as an exposure mechanism of modulation. Cohen's analysis of the biopolitic helpfully articulates the need for new vocabularies to help surface modulation practices and logics. The core legislative and regulatory constructs of existing information privacy reinforce a focus on process and the consequent emphasis on transactional operation. Chapter 8 argued that in doing so, the technocratic nature of information privacy law's semantic base has the effect of reaffirming the importance of process at the expense of providing interruptions against modulation. Further critical questioning of information privacy law's regulatory building blocks, particularly its founding vocabulary, is required, as this will assist discourse about how future forms of information privacy law could develop separate to control and process orientations. A new legal vocabulary would therefore be an interruption to modulation's knowledge production mode, which requires individuals to be left generally unaware of data collection and its resultant purposes for social shaping.

The second and third reforms directly relate to the application of existing information privacy law, and two foundational issues are addressed. The first reform considers whether information privacy law should still be based on protected categories of information – for example, the types of personal information, personal data and personally identifiable information discussed in previous chapters. The second reform regards whether there should still be a requirement for a collection principle as part of the life-cycle protections of personal information exchange. Both act as important interruptions. Relational forms of personal information assist to interrupt the seamless application of modulation because this extends the scope of information privacy law beyond process points and into interstitial spaces and gaps. Similarly, a greater emphasis on fair collections provides a seamful interruption of seamless data collections, particularly those involving sensor data, and interrupts modulation's prescriptive governance mode.

The final reform considers information privacy law in action rather than being an amendment of existing legislative frameworks. Cohen's work on modulation has a significant institutional component, especially regarding the design, construction and implementation of sensorised data collection devices. Chapter 8 detailed the important link between sensor data and two-way flows of modulated practice. Given that information privacy law interruptions of modulation will take place in the gaps and spaces at the outskirts of process points, the institutional endeavours of data collectors that focus overtly on compliance with information privacy law principles would seem misplaced. As such, a shift in institutional practice is required that focuses more heavily on the articulation and implementation of privacy values over information privacy compliance activities, particularly in relation to sensor collections direct from the home. Institutional value setting flows from the three foregoing suggested reforms because it requires critical and thoughtful articulation of an institution's value of privacy. Institutional thoughtfulness exposes the unquestioned logics of modulation and therefore interrupts its knowledge production and governance modes.

In considering these reforms, the relationship between different jurisdictional constructions of protected information and their ability to respond to a power-related context will be critically examined. A parallel will be drawn, utilising the research conducted in the early parts of the book on smart home insurance, between broader constructs of personal data in the European context and the enhanced ability to extend protections to broader notions of informational relationship that move beyond individual control. Similarly, the scope of the collection principles will also be examined from the perspective of a relational context and the power elements outlined earlier to address the question of whether the collection-related principles are still required. It will be argued that they are, because robust collection principles would be an important interruption of modulation, particularly in the sensorised context of the collected world.

9.2   Some Design Points for Future Legal Reform

The design points to aid future legal reform, detailed briefly here, are based predominantly on the transitional move from information privacy law as an individual control-based process to an interrupter of modulation. The points emerge from Chapter 8's critique of the control model's five intended outcomes by cartographically redefining information privacy law's boundary landscapes, based on Cohen's work. These new landscapes start to map information privacy law's terrain as the start of a journey that seeks to transit legal emphases from process to gaps, from seamlessness to seamfulness and from states of unawareness to states of modulation exposure. The design points are thus connectors. They facilitate the mapping of Cohen's core conceptual components to practical information privacy law reforms. As already noted, the suggested design points are not exhaustive, but they should encapsulate some of the major changes required, as evident by the focus on interstitial gaps and spaces.

9.2.1   Interstitial Gaps and Spaces

Chapter 8 detailed how situated intersubjectivity emerges and the need for new forms of boundary management that requires information privacy law to have a broader application focus beyond the process points of principled protections of personal information exchange. If, as Cohen suggests, information privacy has an integral role in protecting selfhood development, and that development unfolds in interstitial gaps and spaces in information exchange processes, then information privacy law will need to intentionally apply in those gaps and spaces also. This suggests that information privacy law's application can no longer solely be fixed on the provision of limited individual rights of control and access aimed at specific points of personal information exchange processes.

Interstitial flexibility, formulated within information privacy law, is needed. As noted previously, the current legislative structures of information privacy law reflect its primary aim: namely, to regulate the process of personal information exchange. New legal vocabularies are required to reflect the broader aims of interrupting modulation and to expose the power-related consequences of the collected world. New vocabularies could then better reflect the core concerns of new conceptual application and could be better suited to the development of new legal definitions that accord greater levels of flexibility in process application. As covered in this section, a core definitional issue would regard the type of information that would give rise to information privacy law coverage that is currently categorised as personal data, personal information or PII.

It could be argued that information privacy law should not employ a regulatory trigger based on certain categories of information, if it is to have a broader role in the protection of interstitial process gaps and spaces. That may well be an argument to consider further, but in keeping with the discussion here and in Chapter 8, the book's analysis will work from the basis that the current information privacy law regime is here to stay and will thus consider evolutionary forms that can unfold in subsequent development. Consideration will be given to what category of information would be more suited to application in process gaps rather than considering the more extensive idea of an information privacy law without an information trigger.[2] Given that the issue of information categorisation is a threshold one, it makes sense to start with that key issue and examine which, if any, of the current constructions would give rise to judicial and regulatory flexible interpretation that could apply in the interstitial spaces and gaps of information exchange processes.

9.2.2   Seamful Interruptions

The second design point regards a stronger role for information privacy law as a specific interrupter of modulation practices and logics. Information privacy law interruptions would be deliberately obstructive legal mechanisms designed to prohibit, slow down or disjoint the

---

[2] Nadezhda Purtova, 'The Law of Everything: Broad Concept of Personal Data and Future of EU Data Protection Law' (2018) 10 Law, Innovation and Technology 40.

seamless data exchanges that can attenuate two-way flows of modulation. To a certain extent, information privacy law already undertakes this role, particularly in relation to the clash of logics between information privacy and big data.[3] Information privacy sets limits on what data-collecting organisations can do with personal information and restricts what data should be collected. There is a set of underpinning logics to the law, as outlined in Chapters 5 and 6. However, the processes of predictive data analytics and big data have a different group of logics. The justification of big data is that by aggregating different datasets, it becomes possible to identify new patterns that provide an unintuitive insight. Any data, collected from anywhere, must be used and kept forever to effectively prospect for data insight.

Chapter 8 suggested that information privacy law's role as an inter-rupter of seamless flow needs to pay greater attention to its disruptive purpose. Quite often, this purpose is used as a perennial criticism about the law's redundancy and its inhibitor of truly innovative data solutions.[4] The segmentation of the personal information exchange process, and the principled protections applied to segments, intends to create a life-cycle protection that acts as interruption. However, the effect of other aspects of information privacy law's construction and application – most notably, its transactional mode of operation and its balancing requirement – have a countervailing effect on its ability to interrupt modulated flow successfully. As noted in Chapter 8, the process spaces of information privacy's transaction mode are continually being crushed by the increasing ubiquity of data collection and analysis. As such, the transactional mode encourages focus on process but at the same time assists in its own crushing by not being able to sufficiently protect against modulated practices and logics of seamless collection, prediction and prescription. Information privacy law's balancing mechanism also contributes to the law's inability to create strong interruption measures. Data collector requirements need to be considered against those of individuals and against economic conditions that promote increasingly seamless exchanges of personal information.[5]

The second guiding point therefore seeks to move information privacy law from self-defeating attempts at seamless balancing of data exchange

---

[3] Yeung, '"Hypernudge"' 126.

[4] For an early articulation of such sentiment see Solveig Singleton, 'Innovation versus Privacy' (Cato Institute 1999) <www.cato.org/publications/commentary/innovation-versus-privacy> accessed 20 June 2019.

[5] van Hoboken 236 with reference to Article 7(f) GDPR, which provides 'particular flexibility' for the private sector.

processes to seamful interruptions of data flows. Attention also needs to be paid to data collection interruptions given the intrinsic link between sensor data collections and ultimate prescriptive outcomes, as detailed throughout the book. Information privacy law requires an augmented collection principle, predominantly based on fairness and lawfulness, to be applied robustly to provide greater boundary-building options for individuals regarding the use of sensor-based collections, especially in sensitive environments such as the home. Such collection principles, of course, already exist, but they should have greater application and consideration as a seamful interrupter that is able to also apply in interstitial gaps and spaces.

9.2.3   Boundary-Building Options

Chapter 8 also noted the conceptual differences between the control model's focus on transactional operation and Cohen's argument for the greater provisioning of boundary management spaces. As Chapter 3 noted, boundary management becomes crucial in collected spaces that are boundary weakened or confused. Information privacy law's interruption contribution would thus be to afford individuals the opportunity of building boundaries around data collection, particularly down to the sensor level. Sensorised devices can be replete with different sensors that have several different collection purposes. Chapter 4's coverage of Nest and Canary devices is a good case study, as a range of different environmental data can be collected in a range of different formats from several sensors.

A meaningful data collection interrupter may require the option of going down to the individual sensor to stop it or delay it from collecting data. Cohen strongly notes that social shaping takes place through our interactions with devices. We are concurrently using these devices and being shaped by them as device interactions shape our understanding of the world around us. All of this suggests that data collection boundaries afforded to individuals would primarily be a matter of legal, institutional and technical design. That would also be in keeping with Cohen's argument that legal application in collected spaces is not purely of the 'defensive and ameliorative' non-interference variety that pervades in attempts to ensure that private spaces become power vacuums. Instead, it requires legal incentives for better, more thoughtful institutional design of afforded boundaries in technology construction.

All the design points, therefore, have an institutional consequence that requires a shift in focus from information privacy compliance of process to the articulation and implementation of internal privacy values that found technical affordances in design. As noted in Chapter 7, this point is crucial regarding the playful and thoughtful development of new

interrupting practices against modulation. The building of internal, institutional privacy values is also more reflective of Cohen's consideration of
social shaping, for which we all take responsibility for. Value-augmented
design, in this respect, admits to the failure of the privacy vacuum and
further accepts responsibility in the provision of seamful collection interruptions as a 'common good' application against modulation. However,
as detailed later, this is a much harder task for data-collecting institutions
who are already bound by their own information privacy law compliance
orientations and who bear legal risk for current compliance failures.

The coverage here regards the need for a much greater value focus for
data collection institutions, but there is not enough space to consider
the role of information privacy law in incentivising these challenging
discussions and design outputs. However, if Cohen is right, that
information privacy is a matter of design, then how information privacy
law encourages the type of critical institutional dialogue required to
afford meaningful boundary interruptions is of importance, especially
as the earlier chapters demonstrate the difficulties involved in control
or ownership-based models in the smart home.

9.2.4   Exposure Mechanisms

The final design point regards a role for information privacy law as a
means for exposing modulation. Cohen identifies a role for information
privacy law as standing in a place between 'truth and power', because
code and law have 'become tools for structuring contests over the material conditions of understanding, participation and self-determination'.[6]
Information privacy law's role consequently has an important position
to play in exposing modulating practices and logics. The exposure element of a reformulated information privacy law would be an important
consideration in the development of new vocabularies to better align the
law with a more explicit power-related role.

It should be acknowledged, as with the other design points, that none
of this will be easy to achieve legislatively, because drafting of information privacy law can prove to be a rather torturous exercise. For example, the final agreed drafting of the GDPR's text took over four years
and resulted in numerous changes at the behest of different bodies.[7]
Similarly, the Australian Law Reform Commission undertook a wide-
ranging inquiry in the operation of the Privacy Act which culminated

---

[6] Cohen, 'The Networked Self in the Modulated Society' 58.
[7] 'The History of the General Data Protection Regulation' (European Data Protection
    Supervisor 2019) <https://edps.europa.eu/data-protection/data-protection/legislation/
    history-general-data-protection-regulation_en> accessed 20 June 2019.

in 2008 and put forward two tranches of reform. To date, only the first tranche, albeit substantial, has been implemented, in 2014, some six years after the report was first delivered, and it is not clear whether the second tranche will ever be considered. That second tranche did, however, include a mandatory data breach notification scheme which has been implemented separately to the reforms proposed.

The consternation generated by the prospect of complicated and fractious legislative amendment, or executive inaction, is not enough reason to preclude critical examination of information privacy's current and future role, particularly from an academic perspective. Nor is the role of exposure a radical suggestion. Several scholars have detailed complex work on technological due process that has suggested a range of different transparency solutions. These include enhanced use of meaningful notification strategies,[8] providing a transparent foundation for predictive segmentation and prescriptive outcome strategies,[9] enhanced and available data-cleansing processes that minimise the risk of inaccuracy[10] and proficient de-identification structures.[11]

To a certain extent, the same could also be said for the GDPR's new right to explanation. Putting aside the debate as to whether it exists or not,[12] the GDPR envisages further protections for individuals in relation to automated decision-making processes that employ machine learning or artificial intelligence, as detailed in Recital 71 of the GDPR. Recital 71 covers a lot of ground and is representative of the fact that a right to explanation encapsulates several GDPR articles. However, it does add an important new transparency and control mechanism by (a) ensuring that individuals can access information on how an automated decision was made and (b) providing an indirect right to opt out of solely automated decision-making systems.

---

[8] Danielle Keats Citron, 'Technological Due Process' (2007) 85 Washington University Law Review 1249.

[9] Frank Pasquale, 'Restoring Transparency to Automated Authority' (2011) 9 Journal on Telecommunications and High Technology Law 235; Tal Zarsky, 'Transparent Predictions' [2013] University of Illinois Law Review 1503; Schermer.

[10] Kate Crawford and Jason Schultz, 'Big Data and Due Process: Toward a Framework to Redress Predictive Privacy Harms' (2014) 55 Boston College Law Review 93, 123.

[11] Ira S Rubinstein, Ronald D Lee and Paul M Schwartz, 'Data Mining and Internet Profiling: Emerging Regulatory and Technological Approaches' (2008) 75 The University of Chicago Law Review 261; Faisal Kamiran and Toon Calders, 'Data Preprocessing Techniques for Classification without Discrimination' (2012) 33 Knowledge and Information Systems 1.

[12] Lillian Edwards and Michael Veale, 'Slave to the Algorithm? Why a "Right to Explanation" Is Probably Not the Remedy You Are Looking For' (2018) 16 Duke Law and Technology Review 18.

In other words, a corporation that utilises solely automated decisions must be able to account for decision-making mechanisms upon request by an individual. If the individual does not like the decision-making process, then they can opt out of the process, if it is solely automated – namely, if there is no human involvement in the decision-making mechanism. The cluster of articles can be broken into two substantive areas. Articles 13 and 14 require notification of decision making, and Article 15 provides individuals with access to such information. All three articles are thus concerned with the provision of meaningful information regarding automated decision making across several different contexts as an improved transparency and fairness mechanism.

The GDPR's right to explanation is a transparency mechanism, but it is one that is still heavily linked to individual control, as explanation is used to form a rational individual decision about an objection to automated processing. In that sense, it still relates strongly to the information disclosure outcome of the control model which Chapter 8 outlined as problematic, particularly in the context of autonomy enhancement and power vacuum creation. The right to explanation is therefore a transparency mechanism that seeks to enhance autonomous decision making based on the provision of meaningful information to obviate the type of cognitive concerns Solove raised about privacy self-management. Consequently, it still has limits as an exposure of modulation mechanism, because it still relies on the precepts of the control model. That said, however, like the brief discussion on academic perspectives of technological due process earlier, it is important to note that there are current transparency initiatives in place that could be developed further into exposure mechanisms. In keeping with the evolutionary development sought, Section 9.3 will examine what reform changes of existing information privacy law structures could be undertaken to move the law towards a specific interruption role. Before we do this, it is first necessary to briefly remind ourselves of Cohen's principles of semantic discontinuity and operational accountability.

9.3   Information Privacy Law Interruptions

Chapter 8 covered Cohen's semantic discontinuity and highlighted its fundamental role as an interrupter of modulation. However, it is also important to consider its role in implementation, especially in information privacy law regimes. Cohen notes that the ability to operationalise semantic discontinuity at scale 'requires multiple and overlapping sets of broadly distributed strategies for durably interrupting (rather than temporarily disrupting) networked information flows'.[13] Besides the need for interruption, which has already been covered in depth, two further elements are notable. First, implementation of semantic discontinuity principles regards multiple and overlapping strategies that are intended to have a broad effect. The design points mentioned previously are based on this element, as they continually overlap in coverage and are intended to have effect across legal, institutional and technical spectrums. Second, the designated interruptions should be durable rather than temporary. Semantic discontinuity implementations should aim to have a temporal and resilient impact. Any implementation of semantic discontinuity principles should therefore be able to persist against the constant two-way flow of modulation but also be flexible to further expand application into newly developing interstitial gaps and spaces. Long-lasting flexibility in application would thus appear key to successful implementations of semantic discontinuity principles, which again suggests a decoupling of scope from process points.

The flexible application of new information privacy law protections would also need support from Cohen's principle of operational accountability, because it is directly implicated in disruptions of modulation. Cohen argues that part of modulation's power regards its ability to act at scale but also to act with very fine precision. This is supported by processes of 'operational opacity' that keep secret the 'cadre of initiates' who exert the wheels of power behind modulation.[14] Operational accountability requires 'meaningful access' to the logics of modulated segmentation, identification of the 'cadre of initiates' and how those logics and individuals present a shaped vision of the collected world.[15] While there are access and transparency mechanisms in current versions of information privacy law, such as the GDPR's right to explanation just mentioned or APP 12's access principle, Cohen's notion of access in terms of operational accountability is much broader.

Its breadth lies in the detachment from information privacy's control base and its refocus on accountability as essentially an affordance protection against modulated forms of social shaping. In other words, a signal virtue is its 'stronger orientation toward practice and toward design'.[16] Access, in this regard, is not about a form of information disclosure mechanism such as the right to explanation, existing access

---

[13] Cohen, 'Turning Privacy Inside Out' 28.
[14] Cohen, 'The Networked Self in the Modulated Society' 77.
[15] Ibid.
[16] Cohen, 'Turning Privacy Inside Out' 21.

principles or privacy policies. Instead, it is about the requirement for institutional affordance dialogues about information privacy interrupters as essential design components and public availability of those institutional dialogues as a means of interrogating the logics, practices and threats of the seamless and invisible infrastructures of the collected world.

The principles of semantic discontinuity and operational accountability disrupt the rhythms of unthinking institutional data collection stupor from which modulation seamlessly forms and flows.[17] New reforms are therefore required, beginning with new sets of legal vocabularies to jolt institutional stupor and to meaningfully aid the access requirements of operational accountability and the institutional affordance dialogues of semantic discontinuity.

9.3.1   New Legal Vocabularies

If modulation's ultimate power act regards the two-way flow of data collection and analysis that facilitates both a conscripted mode of knowledge generation and a mode of prescriptive governance, then exposure mechanisms of two-way flow practices and logics are essential as a counteractive measure of forms of power over and symbolic construction. Information privacy law could clearly have a role as an exposure mechanism, given its coalface application against forms of modulation. Accordingly, how information privacy law is constructed – both in terms of structural, remedial and definitional application – will have a significant effect on its ability to expose seamless and infrastructural indelible practices of modulation.

However, as noted previously here and in earlier chapters, information privacy law's largely tacit role of power amelioration has meant that legislative constructions have focused on giving effect to process protections as part of a transactional mode of operation. The effect of this, as noted earlier, is that information privacy law has a countervailing core that militates from its own ability to prevent modulation. If we accept that the law has a role in interrupting modulation, then it would seem essential that its core definitional and remedial structure adapt to have an explicit power-related role. Exposure would thus interrupt modulation's knowledge production mode, which utilises seamless exchange as a way of fostering industrialised forms of unawareness that privacy policies cannot possibly cope with as an information disclosure mechanism.[18]

---

[17] Ibid 24.
[18] Yeung, '"Hypernudge"' 127.

Cohen's work in the biopolitical domain highlights the effect that different definitional constructs can have on our understandings of modulation viewed from an information privacy perspective. Chapter 8 detailed Cohen's considerations of 'data doubles' and their vital role in establishing the biopolitic from which modulation is protected and grows. Cohen's use of language to describe the biopolitic is helpful, because it exposes the industrialised context of data acquisition in which data from one sensor is inexorably inhered to data from many other sensors. Take, for example, the following referent replacements used by Cohen:[19] data double instead of data subject, data harvesting instead of data processing, data refineries instead of data processers or collectors, data markets instead of personal information exchange processes.

As noted in Chapter 8, this intended use of a critical referential construct assists to expose informational capitalism's core mechanisms of data extraction, agribusiness-type surveillance and industrialised modes of knowledge production. All of this leads to Cohen's notion of trade in the biopolitical domain, where data doubles are herded into segmented groups as the digital equivalent of cattle heading for sale at markets, both real and commodity futures based.[20] Corn-milling processes are even used to expose normalised biopolitical data output as industrialised secondary purposing.[21] In these circumstances, the cultivating of data doubles is always going to be an act of secondary disclosure, because the use of such data will always serve purposes other than what it is collected for. As Chapter 8 noted, the biopolitical attempt to reshape the notion of the common public as a space of private prospecting exposes the governance and shaping power dynamics at play.[22] All spaces, whether public or private, become data harvestable. The paragon idyll of liberalism, the home, is thus nothing more than an allotment to be enclosed once it becomes smart.

Information privacy law's conceptual focus of individual control in processes of personal information exchange simply cannot semantically reflect the industrialised patterns of modulation that emanate from the biopolitic. Its power-related considerations are so tacitly buried within its technocratic construction that it provides minimal leeway for broader structural interpretation, especially in the absence of an underpinning rights-based framework such as that of Australia

---

[19] Cohen, 'The Biopolitical Public Domain'.
[20] Ibid 228.
[21] Ibid 227.
[22] Nissenbaum, 'Must Privacy Give Way to Use Regulation?' 19, and note the helpful discussion of 'we' as the beneficiaries of big data.

or the United States. Even information privacy law in the EU, which, as Chapter 6 demonstrated, affords broader scope for judicial interpretation as a fundamental right of EU citizenship, cannot escape its foundational roots of process control. For example, recent judicial attempts to broaden the scope of a data controller to also include data subjects – namely, individuals as a controller of data processing – are starting to lead to conceptual confusion about the processing responsibilities of joint centralised controllers and data subject controllers, especially in controlling ambiguous spaces such as the smart home.[23]

Some of these confusions and concerns were also surfaced in Part I's coverage of the smart home anatomy and smart home insurance models. The Platform Entity model clearly has a power-related component that regards the ability of any one corporation to set the conditions of entry and existence for all other corporate entities or individuals involved in the smart home space. Being able to set the conditions has a modulated impact that goes down to the level of sensor collections for knowledge production and returns to device affordances regarding the triggering of modes of activity governance. None of this can currently be captured by the language of information privacy law, because infrastructural power is ground down to issues of information disclosure and consent or primary purposes of collection and secondary uses.[24] These are important issues, because if Chapter 4's analysis of the Platform Entity model is accurate, its wide-scale implementation could have significant impact upon individual and collective forms of privacy, which will not receive the type of multifaceted regulatory and judicial attention required.

All of this leads to the need to define what forms of modulated harm a reformulated information privacy law should seek to protect against. That complex issue is not for the confines of this book, but the foregoing coverage seems to make clear that information privacy law's dominant harm construct, the ability of data collectors to identify individuals, will also need to change. Identification, re-identification and de-identification harms do not reflect the legal challenge that modulation poses, because it structurally bypasses these previously important issues. If the biopolitic does proceed to eventual fulfilment, retaining control over the ability to identify becomes redundant because every

---

[23] Lilian Edwards and others, 'Data Subjects as Data Controllers: A Fashion(able) Concept?' (*Internet Policy Review*, 2019) <https://policyreview.info/articles/news/data-subjects-data-controllers-fashionable-concept/1400> accessed 20 June 2019.

[24] Mark Burdon and Mark Andrejevic, 'Big Data in the Sensor Society' in Hamid Ekbia, Cassidy Sugimoto and Michael Mattiolli (eds), *Big Data Is Not a Monolith* (MIT Press 2016).

piece of data can be prospected. It is clear therefore that the construct of personal information should be revisited as a first-step reform of definitional and vocabulary development.

9.3.2   Relational Forms of Personal Information

Cohen outlines the fundamental relationship between personal information and modulation processes. Modulation practices seek to use personal information to shape the prescribed choice preferences as part of its governance mode and a way to understand the behavioural patterns of individuals as part of its knowledge generation mode. As such, the two-way flow of modulation is dependent upon concomitant two-way flows of personal information that provide choices of service, such as those in the smart home insurance models of Chapter 4, and with them political choices of 'information sources, facts, theories and opinions'.[25] The collection of personal information fuels analytic insight and produces prescriptive response. It is no surprise, therefore, that the biopolitical domain asserts a collective ownership of 'raw' personal information that is freely available for extraction. A greater focus on relational forms of personal information would be a counterpoint to the biopolitical domain and would help to interrupt the seamless application of modulation because it would assert cover of information privacy law into the interstitial spaces and gaps that mark that domain.

The problem with the biopolitic, as noted in Chapter 8, is that if there is no boundary between personal and public domains, then everything that is personal is public and everything that is public is personal. Again, this observation warns of the Platform Entity model, where all the different weaves of smart home sensor data are seamlessly capable of being woven together at a central point of collection ubiquity – namely, the voice assistant connectors of devices Google Home or Amazon Echo. Cohen's biopolitic, as potentially manifesting in the Platform Entity model, thus warns of the future modulated threats that arise when the platform giants turn personal information prospection into seamless extraction. Once everything that is personal becomes public, and therefore collectable, it becomes harder for any individual or institution to speak truth to ubiquitous collection powers that produce modulated forms of knowledge production and prescriptive governance. Again, the condition-setting element of the Platform Entity model harks to this concern, because the capacity for shaping entry to smart home markets and how those markets function reflects two-way flows of modulated power.

---

[25] Cohen, 'Between Truth and Power' 65.

Condition setting refers to first-dimension forms of 'power over' – namely, by controlling conditions for market entry. Third-dimension forms of symbolic shaping also exist – namely, the collection powers of the platform suffice to shape understanding of how the market functions and what we should expect from that market.

Cohen strongly observes that personal information collection activities 'are structured by basic judgments about what to collect, what units of measurement to use, and what formats and codings will be used to store and mark the data that are collected'.[26] These observations reinforce that the two-way flow of modulation feeds back to shaping strategies at the data collection device level. Consequently, the macro infrastructures of the biopolitic are unavoidably linked to the micro sensor data collection activities of the billions of sensors that make up the collected world. The fusing of macro infrastructure and micro sensor collections assist to crush the boundaries between personal and public, which means that personal information is everywhere and available to be prospected, extracted and refined.

The power of the Google Home and the Amazon Echo is in their ability to digitally enclose spaces and personal information. 'Enclosure' here is used in the sense provided by Mark Andrejevic as 'the interactive embrace of networked devices that record everything that takes place upon them – they can be captured, stored, sorted, and processed'.[27] But it can also be used in the sense of the enclosure movement of the seventeenth and eighteenth centuries in Britain, where vast tracks of smaller landholdings were consolidated into larger ones in order to produce purportedly more economically efficient modes of agricultural production.[28] Both processes of enclosure create the conditions for two-way flows of modulation from device collection capacities to states of political economy, supported by legal structures. Both also have the same effect of transforming personal information into a public commodity that can then be commercialised for private gain. The centre point of enclosure in the smart home is becoming the voice assistant, due to its ability to enclose, and thus subsume, the data generation capabilities of all other resident sensorised devices, as noted in Chapters 3 and 4. Again, the link between sensor collection and condition-setting capabilities are inseparable.

---

[26] Cohen, 'The Biopolitical Public Domain' 226.
[27] Mark Andrejevic, *iSpy: Power and Surveillance in the Interactive Era* (University of Kansas Press 2007) 2.
[28] JR Whitson, 'Surveillance and Democracy in the Digital Enclosure' in Kevin D Haggerty and Minas Samatas (eds), *Surveillance and Democracy* (Routledge, Taylor & Francis Group 2010) 234.

Given the conjoined relationship of sensor data and condition setting as laying the grounds for modulation to emerge, then it seems clear that significant regulatory attention should interrogate the nature and definition of personal information. Despite the biopolitical claim, personal information is never raw, as it is always 'elicited in carefully standardized fashion'.[29] In other words, data collection of personal information is an affordance of device design. All these considerations highlight a deep relational context to the generation of personal information that goes beyond the direct relationship between individuals, or data subjects, and data collectors, or controllers.[30] The two-way flow of modulation means that personal information, particularly sensor-derived personal information, is refined by its collection context and is therefore referentially related to the broader biopolitic. Consequently, 'relational' in this sense regards the power relations that are involved in the construction of data collection technologies and are intended to extract data for commercial use.

This discussion highlights that information privacy law's threshold issue, of what data triggers regulatory response, must be construed from a broadly relational sense that fully incorporates the power context of the biopolitical sphere. If Cohen is correct in arguing that information privacy's purpose is to protect the interstitial gaps and spaces for playful selfhood to emerge, and information privacy law is to retain its threshold trigger, then the definition of personal information must be capable of being applied at the outskirts, or even outside, of the process of personal information exchange.

Out of the three classificatory modes outlined in the book, it is clear which one has the structural depth to achieve this type of application: namely, the GDPR's construct of personal data. Chapter 6 noted that personal data intendedly has a strong relational element that is signified in the GDPR itself and by standard-setting regulatory agencies. Judicial interpretations have been equally broad, because the rights-based underpinning of the GDPR provides a platform from which personal data can be construed extensively in application. The relational element means that the GDPR's definition of personal data is more suited to provide a basis for protections against modulation, because it has the semantic flexibility to cover ever-expanding constructs of personal information. As discussed previously, if social shaping takes place everywhere, and information privacy has a fundamental role in the protection of selfhood, then it must have the capacity to be applied to

[29] Cohen, 'The Biopolitical Public Domain' 224.
[30] Schwartz and Peifer 146 and the constitutional subtext of the GDPR as a 'rights talk' proliferator.

sensor data collected from everywhere. The levels of semantic flexibility required to achieve such broad-scale application simply does not exist in the conceptual basis of personal information in an 'about' sense, or as PII in the United States.

In fact, the weaknesses of both Australian and US approaches are evident in recent legislative proposals. The Australian government is seeking to implement a data portability mechanism through a new regime entitled the Consumer Data Right (CDR). It is made up of a complex two-track information privacy regime to allow consumers to port 'CDR data' to different service providers. CDR data, for the benefit of the CDR regime, employs the GDPR definition of personal data. The Explanatory Memorandum states why:

CDR data is data that 'relates' to a CDR consumer. The concept of 'relates to' is a broader concept than information 'about' an identifiable or reasonably identifiable person under the Privacy Act. The term 'relates' has a broader meaning than 'about' and is intended to capture, for example metadata of the type found not to be about an individual in *Privacy Commissioner v Telstra Corporation Ltd.*[31]

The use of 'relates' in the CDR regime text is therefore used to provide more expansive protections than the Privacy Act's definition of personal information, particularly following the Full Federal Court's *Telstra* decision. The definitional use of 'relates' is specifically tied into the rights-based framework of privacy protections in the GDPR and the EU more generally.[32] More data thus falls under the ambit of the GDPR as a policy counterpoint to the increasingly widespread use of data aggregation strategies.[33] The use of 'about' in the Privacy Act was specifically designed to curtail the widespread application of the Act and confine it to a reduced range of information, to create certainty and minimise the regulatory load on regulated entities.[34] It was some

---

[31] Explanatory Materials, Treasury Laws Amendment (Consumer Data Right) Bill 2018 (Exposure Draft, Second Stage) 10.

[32] Regulation (EU) 2016/679 of the European Parliament and of the Council of 27 April 2016 on the protection of natural persons with regard to the processing of personal data and on the free movement of such data, and repealing Directive 95/46/EC (General Data Protection Regulation), Recitals [1]–[3]; Article 29 Data Protection Working Party, Opinion 4/2007 on the Concept of Personal Data, 4.

[33] Article 29 Data Protection Working Party, Opinion 13/2011 on Geolocation Services on Smart Mobile Devices; Article 29 Data Protection Working Party, Opinion 8/2014 on the Recent Developments on the Internet of Things (14/EN WP 223, 2014); European Data Protection Supervisor, *Privacy and Competitiveness in the Age of Big Data: The Interplay between Data Protection, Competition Law and Consumer Protection in the Digital Economy* (European Data Protection Supervisor 2014).

[34] Australian Law Reform Commission 306.

of these general points of statutory interpretation that formed the AAT decision,[35] which appear to receive tacit support in the Federal Court's decision in the *Telstra* case[36] and from which the Australian government now seeks to depart. The 'relates to' aspects of the CDR scheme will likely give rise to problems within the broader Australian information privacy law framework. It will eventually create two separate jurisprudential tracks of reasoning regarding the type of data that will be covered under the CDR scheme and the Privacy Act.

The GDPR's broader definition is also beginning to seep slowly into the legislative frameworks of state-based US laws. The 2018 enactment of the California Consumer Privacy Act, which comes into effect on 1 January 2020, partially departs from the predominant construct of PII, as it provides coverage for 'personal information' which is defined as 'information that identifies, relates to, describes, is capable of being associated with, or could reasonably be linked, directly or indirectly, with a particular consumer or household'.[37] Similarly, in 2019, the newly proposed New York Privacy Act facially adopted the GDPR's definition of personal data as 'information relating to an identified or identifiable person'.[38] However, it should be noted that while the broader scope of the 'relates to' construct is beginning to appear in US state-based law, it does so in a hybrid fashion. Both the Californian law and the New York proposal supplement the breadth of scope that the GDPR definition would allow with the types of specified types of data redolent of the PII model. The significant lists of PII in both developments are not meant to be exhaustive, but they are likely to have the impact of constraining future jurisprudential development to certain categories of information.[39]

Hybrid imbuement of the GDPR construct of personal data into Australian and US information privacy law frameworks is a tacit admission that the definitions of personal information and PII do not allow enough flexibility to cover ever-expanding types of sensor data that should give rise to information privacy law coverage. That said, however, significant problems are likely to emerge if the GDPR's definition of personal information is to apply to everything and everywhere, as thoughtfully noted by Nadezhda Purtova, who cautions

---

[35] *Telstra Corporation Limited and Privacy Commissioner* [2015] AATA 991 (18 December 2015), [35], [98].
[36] *The Privacy Commissioner v Telstra Corporation Limited* [2017] FCAFC 4 (19 January 2017, [41].
[37] Section 1798.40(o)(1), The California Consumer Privacy Act 2018.
[38] Section 1100(8), New York Privacy Act Bill 2019.
[39] Solove and Schwartz.

of such a broad application of information privacy protection. Purtova's concerns arise from 'system overload' – namely, that information privacy law was not conceptually intended to have such universal reach and 'the high intensity of data protection compliance' due to the 'highly intensive and non-scalable regime of rights and obligations'.[40] Purtova is therefore not necessarily concerned about a broad interpretation of personal data, but rather the role it plays within the framework of information privacy law, including its compliance expectations.

The circle consequently starts to be squared as the real concern begins to surface. The concern, as Purtova rightly highlights, is not the breadth of personal data's potential application. Instead, it is the concern raised in Chapter 8 of the control model's overt focus on the exchange process of personal information. It is that focus, rather than an expansively constructed definition of personal data, which gives rise to system overload issues. Decoupling information privacy law from this overt focus will not only assist with a broader power-related role, but it will also rebut the often-trotted-out claims of its redundancy in the face of big data logics and practices. Accordingly, as demonstrated throughout this book, the need for information privacy is not redundant or irrelevant. In fact, the discussion of its role in interrupting modulation indicates the opposite. What has changed, however, are the structures of data collection and exchange which extensively challenge information privacy law's control basis. Much greater attention should therefore be placed on the role of information privacy law at the point of data collection, including the strengthening of collection principles based on fairness.

9.3.3   A Greater Emphasis on Fair Collections

A number of issues are covered by this book, but one discussion stands out more than the others. There is little doubt that information privacy law needs to provide some form of protection at the point of data collection.[41] Chapter 6 noted the debate unfolding in academic and regulatory circles about the removal of collection principles and replacing them with a greater focus on 'use' as part of notice-and-consent models. There is a clear jurisdictional disposition to these debates, as the US sectoral approach is structured to propagate the notice-and-consent model at the expense of other information privacy principles. Chapter 6 noted

---

[40] Purtova, 'The Law of Everything' 75.
[41] Nissenbaum, 'Must Privacy Give Way to Use Regulation?' 8 makes clear why this is necessary in the context of broad and narrow definitions of collection.

that decision is itself a political one about the accord given to information privacy law protections in the balancing of data collector and individual provider interests. Chapters 3, 4, 7 and 8 then demonstrate the challenges that the control model, predicated on the privacy policy as an information disclosure mechanism, will face in the collected world, utilising smart home business models as a prime example.

In doing so, the book adds to the chorus of academic criticism about the applicability of privacy policies and notice-and-consent models that underpin information privacy law's control basis. The privacy policy does not operate as an effective information disclosure mechanism, and it underpins a conceptual base that is irrevocably weakened by the onset of the collected world. Put simply, neither can hope to interrupt modulated two-way flows as they currently operate. Neither can cope with the vast volumes of sensor data that make up the collected world. Neither can adequately articulate the challenges that arise from predictive and prescriptive application at scale.

It is time to end the fatuous adherence to the privacy policy as *the* fundamental component of information privacy law's operation, as detailed in Chapter 8. It is not an effective rationality-promoting mechanism to enhance autonomous decision making. It does not create the expected power vacuum for autonomous decision making to flow. It does not support the ability for individuals to create or manage the type of boundary necessary to negotiate the complexity of the collected world. It does not assist with the amelioration of information asymmetries. It does not provide the intended basis for the control model's desired balancing mechanisms.

Instead, its effect is to serve the interests of data collectors who attempt to secure consent[42] through individual unawareness, obsolescence and cognitive fatigue.[43] Once consent is secured, as Chapters 7 and 8 note, then Chapter 4's data exchange models can begin to flow, even though individuals are likely not to appreciate the full consequences of information provision, particularly in models as technologically and structurally complex as the Platform Entity model. The utter complexity of data provision possibilities that arise from sensor collections – intended to capture environmental activities and to create a behavioural 'data double' for commercialisation purposes – is such that we will never be able to formulate a truly rational decision-making capacity. How could anyone ever truly understand that the way we act in the home – such

---

[42] Cohen, 'The Surveillance-Innovation Complex' 218.
[43] Nora A Draper and Joseph Turow, 'The Corporate Cultivation of Digital Resignation' [2019] New Media & Society.

as how we use or do not use a remote controller, change a battery in a smoke sensor, restock a fridge or leave washing in a washing machine – will be used in decisions about a whole range of commercial services that are simply beyond our comprehension?

Information privacy law needs to have designated protections for collection as part of its life-cycle ambit that are capable of application in the interstitial spaces at the edges of, or beyond, exchange processes. A collection principle would be an important application of semantic discontinuity, as it would provide a seamful interruption of seamless data collections, particularly those involving sensor data. A collection principle would specifically aim to interrupt modulation's prescriptive governance mode. In doing so, it would also 'advance the goal of boundary-setting' by providing cognitive breaks that expose the 'irreducible nonhumanness of our automated tools'.[44] In other words, a collection principle assists to expose the invisible tramlines of modulated power that are deeply embedded in device and infrastructural capabilities. Exposing the tramlines also has the effect of exposing the link between individual sensor data collections and the ultimate terminus, the biopolitical domain.

A collection principle would need to serve several overlapping purposes, in keeping with the type of protective structures that emanate from the application of semantic discontinuity and operational accountability. It would need to be seamful to interrupt seamless and continuous data exchanges. It would need to be flexible as it applies in interstitial gaps and spaces at the outskirts of process. It would need to afford the creation and management of boundaries as part of technical and institutional design. It would need to have visible effect as a means of exposing invisible infrastructures of sensor generation, collection and prescriptive use. All of this suggests that a collection principle would need to operate at a higher level of abstraction that goes beyond process point management.

An interruption-focused collection principle is unlikely to be based solely on the application of minimality and purpose specification principles that currently find greater favour in comprehensive information privacy laws such as the GDPR or the Privacy Act. Both the minimality and purpose specification principles are so heavily welded to process points that they do not provide the type of interstitial flexibility required. That is not to say that they provide no flexibility or lack importance. Far from it. Setting boundaries about the type and level of personal information that could be collected to meet a specified

---

[44] Cohen, 'Turning Privacy Inside Out' 29.

business purpose and use will provide inbuilt boundaries and could provide interstitial coverage in gaps. For example, the very notion of limiting the scope of collected information has the effect of creating a barrier that prevents data ubiquity. The minimality principle has the effect of parsing personal information into flows based on necessary and unnecessary categories. The former become part of the exchange process, and the latter should not be collected if it is not related and not required for a specific business purpose.

At this point the privacy policy, again, becomes crucial because it can legitimate broader collections of personal information for equally broad construction of business purposes, once consent for personal information provision is secured.[45] However, a problem emerges in sensorised spaces such as the smart home, because, as Chapter 7 detailed, data collection is prospective in nature. Smart home data collectors know that their pot of gold lies in the prospecting of behavioural patterns, but they do not fully comprehend at the point of initial collection how their business model will fully operate for valuable insight to emerge. The business model is dependent upon collections of sensor data used to form personalisation outcomes, which means that commercial value is emergent, evolutionary and always subject to change. The latter point is important because personalisation outcomes will alter in tandem with behavioural changes that emanate through sensor-based data doubles. In these circumstances, the scope of personal information collection must be broad enough to encompass the nascent qualities of sensor-based analysis that seek to produce valuable insight about behaviours. The purpose of collection must be construed broadly, which has the corollary effect of weakening minimisation capabilities.

These considerations re-emphasise the need for broad interpretations of relational personal data, as both minimality and purpose specification principles will only operate when collected data is classified sufficiently to trigger regulatory or legislative coverage. The reliance on information classification to activate minimality and purpose specification principles again suggests that while both are assistive as boundary-producing protections, they have limits, especially regarding the ability to extend into interstitial gaps and spaces. It is therefore argued that the quieter cousin of minimalisation and purpose specification, namely fairness, should be given greater weight as an interrupter of modulation.[46]

---

[45] Yeung, "'Hypernudge'" 127.
[46] Cate and Mayer-Schönberger note the absence of a discussion about a fairness principle, specific to collection, in roundtables of privacy experts leading to the reformulation of the 2013 OECD Guidelines.

Chapter 6 highlighted that both the GDPR and the OECD frameworks have a fairness component that applies in relation to collection. In the GDPR, it emanates through fairness being a component part of processing. In the Australian Privacy Act, APP 3.5 requires data-collecting entities to not collect personal information unfairly, as well as unlawfully, much the same as does the GDPR. Fairness and lawfulness are therefore key requirements for personal information collection in the EU and Australia. It should be noted that a fairness obligation is generally absent in US sectoral information privacy laws, given the focus on specified categories of data construed as PII and a tendency to favour information disclosure mechanisms at the point of collection. As such, the absence or inclusion of fairness as a specific requirement of collection again relates to different political motives for information privacy law protections.

Cohen states that the 'power of modulation derives partly from its precision, but partly from its operational opacity'.[47] Fairness, as an element of semantic discontinuity, provides the means of slowing, stopping or dispersing the precise nature of data collection via sensorised devices. Fairness, as an element of operational accountability, sheds relatable light on modulation's deliberate opacity. Fairness is the thread that binds the seams of seamful interruption.

A robustly supported fairness principle, designed in law for accompaniment in technical design, has the broader stroke of application that can interrupt forms of 'power over' or as omnipotent social shaping. Moreover, in terms of operational accountability, the application of a fairness principle at the point of collection could provide a greater level of transparency that goes beyond the information disclosure aim of a privacy policy. The type of transparency envisaged here is heavily implicated in technical design that intentionally seeks to interrupt seamless collections of personal information. Most importantly, it is conceptually flexible enough to spread from process into gaps and spaces, and it could therefore afford the management of boundaries as part of technical design.

Take, for example, the construction of Australian Privacy Act's collection principle, APP 3.5, which requires collections of personal information to be lawful and fair. Chapter 6 detailed that a fair collection does not involve intimidation or deception and is one that is not unreasonably intrusive and is inherently contextual.[48] In other words,

---

47 Cohen, 'What Privacy Is For' 1932.
48 Commissioner, *Australian Privacy Principles Guidelines* 14.

it depends on the circumstances of each collection. Context is an important consideration in relation to fairness. For example, generally, a covert collection will be seen as one that involves deception, so it is more likely to be unfair.[49] However, some covert collections in certain contexts may be required, such as in benefit or fraud investigations, where to notify the individual about the collection would defeat its purpose.[50] In these circumstances, it could be reasonable for a seemingly unfair collection to be deemed fair.

The application of unfairness as the basis for a collection principle has important consequences, because it makes it easier to extend regulatory application beyond process. Fairness – as deception, intimidation, covertness or unreasonable intrusiveness – extends the regulatory focus beyond information exchange process points to broadly consider the actions of data collectors as part of their relations with individual personal information providers. Fairness, as applied at the point of collection, is consequently relational in focus. The principle ascribes legally enforceable expectations of relational acts at the point of collection. As such, it adds a different dimension to the application of other collection principles, such as minimality and purpose specification. The two latter principles, as noted previously, are intended to minimise the amount of personal information collected by relating it to data collector operations involving specified and justifiable uses of that information. Both principles consider information privacy obligations as part of process. The fairness principle considers information privacy obligations as relational acts.

The consideration of relational acts of unfairness has important consequences regarding sensor collections, such as those in the smart home. Those collections require sensors to be deeply embedded into technology so that they are unnoticed by environmental occupants. Unawareness is a key requirement for the purported richness of sensor data, because it enables the collection of data from individuals at their most natural and raw. The consequence of the fairness principle is that sensor data collections can be examined from a different and broader information privacy context – namely, the actions of data collectors in relation to collections from individuals. This context opens a range of different actions that go beyond the confines of process and privacy policies.

For example, Chapter 7 demonstrated the importance of situating sensors in smart environments, because the location of the sensor will have a major impact upon what data is collected and from whom.

---

[49] Ibid.
[50] *Griffiths v Rose* [2011] FCA 30 (31 January 2011).

A fair collection in spaces such as the smart home may therefore require data collector agency in the location of sensors, to avoid the type of deceptive or unreasonably intrusive collections that would be unfair. How a collector relates to an individual regarding sensor placement opens a range of exposure, boundary management and seamful interruption opportunities for individuals. Unawareness of sensor collections will be challenged through exposure to sensor data collection capacities. Once exposure is gained, then boundaries of different varieties can be playfully created and managed to suit different needs and requirements. The awareness of boundary possibilities and their actualisation leads to seamful interruptions.

All of this is possible because the extended application of fairness can reach into the interstitial gaps and spaces that Cohen encourages us to care for. Cohen also notes that data collection requirements form the basis for an affordance dialogue by institutional data collectors.[51] By simply begging more strongly the question of whether a form of data collection is unfair, it incentivises thoughtful institutional considerations about the information privacy affordances required as part of technical design. In turn, that will require a greater institutional focus on value articulation, particularly regarding information privacy.

9.3.4   Incentivising Institutional Value Exploration

A strength of Cohen's analysis regards its exposure of the underlying values that especially underpin modulation's knowledge production mode. The techniques of modulation are deployed and aligned with a 'deeply internalized' system of values that is 'calculative, instrumental and unreflective'.[52] These deep-seated values often reveal themselves as the unquestionable innovation antithesis of information privacy. Once innovation values are stacked against others, including privacy values, then serious, critical and thoughtful institutional discussion is stymied because 'no one wants to go on record as opposing innovation'.[53] Accordingly, a debilitating institutional effect unfolds where

[c]orporate internalization of the need for privacy oversight does not necessarily result in effective internalization of privacy-related values and imperatives by the technologists who design networked digital products and services.[54]

51   Cohen, 'Turning Privacy Inside Out' 21.
52   Cohen, 'Between Truth and Power' 66.
53   Cohen, 'What Privacy Is For' 1919.
54   Cohen, 'Turning Privacy Inside Out' 26.

Cohen's assertion is particularly important in the context of interruptions to two-way flows of modulation. As noted throughout the last two chapters, effective interruptions of modulation 'go all the way down' to the design, construction and use of the sensor.[55] This would suggest the need for 'a complementary set of principles and practices for rethinking the design of material and technical infrastructures from the ground up'.[56] Going 'all the way down' requires a distinct role for corporate institutions regarding the development of an institutional value of information privacy. If modulation flows seamlessly, and social shaping takes place everywhere, then institutions that are the collection gatekeepers of sensor data have an important role in the creation of interruptions and the promotion of interstitial gaps and spaces. The protective measures of information privacy law are thus matters of institutional design, which requires a major shift in what information privacy law incentivises in corporations.

In other words, meaningful value development should have the same, if not greater, internal consideration as forms of overt process compliance. The institutional focus of data collectors therefore needs to move from avid compliance with legal and technical processes to a much more difficult, challenging and ongoing task – namely, the articulation and pronouncement of value sets, including information privacy. Value articulation will then afford a range of different collection options for individuals that allows them to interrupt the seamless flow of sensor data. Value development and 'going all the way down to the sensor' are intimately linked, because the principle of semantic discontinuity is ultimately a matter of design. All of this suggests an important institutional role for 'tinkering',[57] as outlined in the final chapter. Tinkering in one sense is innovation, but of a different kind that does not seek to maximise profit out of the sensorised data doubles of the collected world. The innovation here is in the thoughtful institutional role of critique. And this thoughtful role is going to be increasingly important when institutional data collectors use sensor collections from the smart home. Institutional thoughtfulness exposes the unquestioned logics that underpin modulation and therefore interrupts its knowledge production and governance modes.

A key theme discussed in this book's smart home research has been the uncertain quest for value generation in smart home environments. Chapter 4 reported on the increasing emphasis of major insurance

[55] Cohen, 'Affording Fundamental Rights' 88.
[56] Ibid.
[57] Cohen, 'What Privacy Is For' 1920.

companies on having a smart home business presence through partnerships with smart home system and device providers. The technological fragmentation and legal uncertainty about regulatory frameworks for smart home development require thoughtful consideration from corporations seeking involvement in this space.

Chapter 7 highlighted that the control basis of information privacy is significantly challenged in the smart home setting because of the different data exchange relationships that pervade. Notions of data ownership, and thus notions of control, are dispersing rapidly in the smart home. These dispersions give rise to the need to establish a defined information privacy value for smart home business model development, given the link between corporate sensor collections and modulation flows. Two broad options essentially emerge for smart home corporate data collectors: data extraction and affordance of design interruptions.

Data extraction regards the increased and sensorised smart home data collections that can generate insight about the behaviours, emotional patterns and moods of smart home participants, especially where sensor data becomes biopolitically extractable due to the diminishing of boundaries. The home is a protected legal space for good reason. We are our true selves in the home because it is a space in which we can conduct selfhood experimentation without fear of social sanction. The home, as noted in Chapter 5, is the prime site for individual autonomous growth in liberal societies based on the ability to build boundaries that protect the privateness of activities in that space. However, Chapter 7 detailed that the smart home is not the private space that it used to be. The smart home seeps data to multiple data collection parties, and this data can be analysed to identify intrinsic participant behaviours. If these behaviours are identifiable, then it can become possible to nudge participants towards certain outcomes that could benefit smart home system or device providers rather than the participant.

If the valued asset of smart home business models is the accumulation of sensor data to form a tradable data double, then the difference between data extraction and participant exploitation becomes increasingly narrowed by modulated processes. If, to put it another way, personal data really is the new oil, then the smart home participant is the oil well to be primed and pumped, with the collector knowing that it will never run dry. Data extraction, therefore, is really about notions of ownership and control over individuals as exemplified by modulation's modes of knowledge production and prescriptive governance. Modulated-focused, smart home business models thus aim to retain and secure the ownership and control of the customer, both in data and corporeal forms.

Smart home business models based on affordances of design interruption undertake the opposite position. As highlighted throughout this book, the collected world challenges notions of ownership and control across several different axes. Ownership and control of smart space sensor data, by any one party, is becoming increasingly problematic. Newly implemented legal frameworks, such as the portability regime of the GDPR and Australia's CDR, attempt to conceive the data relationships between consumers and businesses in a fundamentally different way. Rather than data being owned and controlled as a business asset, it is a joint resource for consumer and commercial service innovation, albeit still grounded in notions of control.

However, the overall regulatory effect is potentially significant. Rather than being the new oil, personal information needs to be the new institutional dialogue as a predicate for a value-based outlook. Affordance-based business models, unlike those of the data extraction variety, thoughtfully acknowledge the complexities of the fragmented environments of the collected world, particularly the smart home, and seek to act in the best interests of the smart home participant by helping them negotiate the true complexities of the biopolitic. Operational accountability, in this sense, is not based on information disclosure mechanisms as privacy policies. Rather, one of the key roles of the data collector is to assist with the development of vocabularies so that customers can understand these complexities and what it means to provide sensor data for prescription. A key design affordance is thus the articulation and presentation of a core institutional value of information privacy. Nudging, therefore, does not consist of automated forms of modulation for purely profit-making purposes. Instead, it becomes an exposure catalyst that enables deeper degrees of trusted dialogue and allows the collector to better advise the participant in the negotiation of complex data environments.

The models of data extraction and affordance of interruptions give rise to foundational questions about the role of information privacy law in collected world business models. Again, two foundational options arise that underpin both models: compliance orientation with process control and value engagement.

A 'tick-in-the-box' compliance orientation reinforces the modulated tendencies of business models based on data extraction. An overt compliance orientation approach towards information privacy is more achievable in traditional forms of personal information exchange involving direct collections of personal information from customers. As detailed in Chapter 7, sensorised collections from the collected world are an entirely different matter. Sensor collections are specifically intended to enable the identification of individual behavioural patterns. This book has shown that these behavioural patterns will involve participant interaction with the home environment and its sensorised devices, but it will also enable the inference of sensitive knowledge that the participant would not want to disclose. Chapter 7 highlighted that participant understanding is crucial if the privacy self-management basis of information privacy law protections is to operate effectively. However, Chapter 8 then showed that the degree of participant understanding about the true complexities and consequences of sensor collections is likely to diminish, given the complex nature of smart home systems.

Current information privacy compliance is achievable by the provision of carefully crafted privacy policies that notify customers about the application of information privacy principles pertinent to the relevant data collection. At the back end, a privacy impact assessment could be conducted which confirms that the application of information privacy principles is in place and can be audited. Both are integral information privacy compliance measures. However, Chapter 8 noted that it is unclear how these measures apply in a system where control is significantly diminished. The expectations of privacy about home activities are such that any breach of those expectations, regardless of whether valid or invalid, will likely give rise to significant reputational risks for smart home data collectors. Privacy in the smart home data collection context cannot solely be considered as an information privacy compliance issue, even though there are advantages of doing so under business models based on data exploitation of smart home event data. Even trying to determine which smart home data would or would not be personal information for listing in a privacy policy would be a challenging and uncertain task, as detailed in Chapter 7.

Information privacy under an affordance model would have a different application that goes beyond solely information privacy considerations. Privacy is considered holistically as interruptions of modulation that go beyond issues of control over personal information. Chapter 7 also highlighted that different participants in the smart home may themselves have different levels of privacy expectations regarding smart home event data. These expectations may also vary regarding different types of collected data (e.g. video and device metadata). All of this militates against the notion of a one-size-fits-all approach to smart home privacy, especially in the form of boilerplate privacy policies and compliance orientation.

Instead, privacy becomes part of the dialogue catalyst and an opportunity to build more deep and meaningful levels of trust with participants through open explanation of the true role of information privacy in biopolitical environments. Open explanation, in a trusted, collected role, is of course incompatible with a model founded on data extraction which ultimately seeks to engender a modulated perspective of privacy expectation that favours business interests over individuals. All of this leads to the ultimate and most difficult question of all, and one that is a challenge to businesses constantly 'disrupted'. That question is *Who am I?*

To answer this question requires an institution to have a clear value of privacy that it stands for – a value of privacy that recognises its role in providing interruptions to modulation and is the enabler of a deeper form of trusted engagement with individuals. If, as is suggested earlier, the ultimate value proposition of collected world business models resides in the customer rather than the data, then business models must be reflective of core values, including a value of privacy, to foster deeper engagement with customers. Developing a meaningful value of privacy that goes beyond information privacy compliance is a significant challenge but one that is clearly required if an institution takes seriously its affordance role.

At the very least, a clearly articulated value of privacy will better prepare an institution for challenges in the future, in which we enter societies where the ubiquitous forms of data collection become the norm. In these societies, it will become increasingly hard to differentiate information privacy compliance from real outcome prescriptions, because the crushing of process will enfold one seamlessly into the other. A value of privacy in these settings fully regards the dangers of modulation which are dependent upon two-way flows of information through institutions. Privacy is not a regulatory roadblock or a redundant irrelevance. Instead, it is a dialogue catalyst that helps expose the dangers of modulation and clearly affords design options of interruption for individuals. Ultimately, it is a dialogue based on equitable acknowledgements of the customer role in building information societies that go beyond traditional notions of owning the customer for extraction purposes. A clearer articulation of institutional privacy values thus interrupts modulation's governance mode by creating tinkering spaces for design affordances to develop and interrupt modulation's knowledge production mode by critically asking the simple questions Who am I? and What do I stand for?

9.4   Conclusion

The traditional information privacy process provides a range of protections across a life-cycle process. Part III of this book has argued that these process protections will need to be supported by others to provide effective protections against modulated practices and logics. The last three chapters, including this one, have critically examined the veracity of information privacy law's principled exposition, as a form of individual control. The chapters have put forward some design points for future legal reform and started a discussion about the type of information privacy law interruptions required to better align the law with an explicit power-related role. That role regards two-way flows that both reach down to individual sensors and up to macro infrastructures.

These changes will require a new role for institutional data collectors, who will need to have clearer understandings of their own values towards customers, privacy and ownership. Value development and tinkering spaces to translate values of privacy into technical design are crucial in the context of power flows that reach up and down. These are the institutional actions that assist to interrupt modulation by becoming part of the design framework of the collected world. Rather than unthinking and inhuman, these types of activities will assist to ensure that the two-way flow of power becomes thoughtful and considerate. These new flows help ameliorate the underlying logic of big data as part of the broader process of unthinking that provides the power basis of modulation. Institutional tinkering and the affordance of playful weaving have important parts to play if we are to stop the transition from a smart to a collected to a modulated world.

10   A Smart, Collected or Modulated World?

Now that my argument is laid out in full, I will take the liberty of wrapping up the book in the first person, with some reflections also from my own research. I was fortunate to see Virginia Eubanks speak in 2019. Eubanks' work,[1] like Julie Cohen's, foundationally challenges the justification and application of automated forms of decision making by examining their effect on individuals. Despite the sheer awfulness of some the prescriptive outcomes she encountered, Eubanks noted strongly at the end of her talk that it was important to end on positive solutions or suggestions; otherwise there is a danger that we reinforce discussion about privacy as part of an unchangeably bleak and dystopian future. That observation struck a chord with me, as I sometimes feel that my own work perhaps asserts the problem more than it does the solution. Positivity, though, as Cohen relates quite beautifully, is essential if we are to ensure that the benefits of smartness which flow from the collected world do not result in modulated forms of power.

In this final chapter, I therefore ask whether we are heading towards a smart, collected or a modulated world. A smart world is characterised by the unerring wonder and seamlessness of devices and infrastructures that can record the world through sensors and the data they collect. The smart world is non-critical and promotes the unbounded wonderment of having knowledge about everything as a presupposed good that will benefit all, equally. The collected world is predominantly descriptive. It provides a view of smartness that strips it of its admiration, to identify and examine the consequences of ubiquitous sensor data generation and the collection, analysis and sense-making infrastructures required to create valuable insight. In so doing, the frame of the collected world assists in exposing the underlying logics at play and the effect of sensorisation, particularly in relation to information privacy law.

---

[1] Virginia Eubanks, *Digital Dead End: Fighting for Social Justice in the Information Age* (MIT Press 2011); Virginia Eubanks, *Automating Inequality: How High-Tech Tools Profile, Police, and Punish the Poor* (1st edn, St. Martin's Press 2018).

The modulated world, as Cohen's work shows, is definitively critical and is intended to surface the deeply embedded contexts of power that flow in modulated forms of knowledge production and prescriptive governance.

The differences between a smart, a collected and a modulated world encapsulate the book's transitions through its three parts. I started in Chapter 1 by asking a question.

> *To what extent does the collected world therefore require us to rethink the conceptual basis of control-based information privacy and its manifestation as the principled protections of information privacy law?*

That question has guided the book's analysis, and its answer has unfolded across a complex journey.

Part I contended that the smart world is really the collected world. Chapter 2 provided an overview of technological development covering three areas – namely, smart individuals, buildings and environments. A key element of all these areas is sensorisation and the rapid spread of sensorised devices that enables new forms of collection. These new forms were examined further in one of the prime sites of sensor data commercialisation, the smart home. Chapter 3 detailed the complex data generation anatomy of the smart home and examined it from its sensing, reasoning and intervening processes. Again, sensorisation is key to the operation of the smart home and the business models that are now starting to develop. Chapter 4 examined smart home commercialisation in greater depth to conceptualise three smart home insurance business models: Partnered Data Acquisition, Partnered Intermediary and Platform Entity models. These models highlighted the commercial uses of home-generated sensor data and the logics, new business relationships and outcomes that flow from it.

Chapter 4's analysis of privacy policies also highlighted some key differences in jurisdictional approaches to information privacy law and gave insight into some of the challenges to be faced in a collected world. Part II, in preparation for a critical examination of these challenges, detailed the conceptual bases of information privacy and its implementation in information privacy law. It then examined what information privacy law seeks to protect and how it provides protections.

Chapter 5 identified four conceptual themes that have underpinned information privacy's development – namely, individual control over personal information, informational access and personal autonomous growth, a broader social and relational context, and privacy as a structural form of power. Chapter 5 highlighted that the control concept is dominant, as evident by its implementation as information privacy law, discussed in Chapter 6. That chapter identified the different foundational structures and jurisdictional perspectives of information privacy involving EU, US and Australian legal frameworks. Two fundamental issues were then detailed: the types of regulated information that trigger regulatory response – namely, personal data, PII and personal information – and information privacy law's principled process of protection. Attention was given to collection principles as a means of outlining foundational differences between sectoral and comprehensive regimes of information privacy, particularly regarding the overt use of a notice-and-consent model.

Part III then examined the consequences of the collected world and the challenges they will bring for information privacy law's dominant control model and its manifestation in information privacy laws predicated on process protections. Chapter 7 returned to the smart home as a way of examining collected challenges. It demonstrated that sensor data collections are different to the type of data collections envisaged by the control model of information privacy. Moreover, the very notion of control is challenged in boundary-dispersed environments like the smart home, which are essentially fragmented and contested. These factors give rise to significant challenges for the control model of information privacy and its focus on the process of personal information exchange.

Chapter 8 then returned to the conceptual analysis of Chapter 5 to address what information privacy should protect in a collected world. A sustained critique of the control model was put forward in relation to five intended outcomes of information privacy law. Cohen's work was then introduced as a means of further critiquing the control model and reshaping a conceptual focus of information privacy based on a more explicit power-related role. The new focus would see a shift in what information privacy seeks to do – namely, transit from autonomy to situated intersubjectivity, power vacuums to modulation, transactional operation to boundary management, information asymmetries to social shaping and balancing mechanisms to expose modulation.

New forms of information privacy law could develop, explicitly designed for the power-related consequences of the collected world. Chapter 9 examined how new forms could develop and highlighted some design points for future legal reform, putting forward some information privacy law reforms which could act as interruptions to modulated power. In doing so, Chapter 9 returned to the question posed in Chapter 1 and to Chapter 6's analysis. It highlighted the need for relational forms of regulated information and a collection principle based on fairness. Both reforms would require support from new legal vocabularies and new ways of incentivising value discourse in data-collecting institutions.

I now want to set out what's at stake if the modulated world does come to full fruition. As such, it is necessary to return to Julie Cohen's work one last time before considering the playful antidotes to modulated power, institutional tinkering and individual weaving.

10.1   The Dangers of a Modulated World

The strength of Cohen's analysis lies in her ability to link the smallest technological activities that take place on the ground and implicate them into much broader societal considerations. The devices and infrastructures that make up the smart and collected worlds can also have a modulating effect, because they increasingly mediate and shape our experiences with the world. This power of mediation means that the very practice of citizenship can be configured in subtle ways.[2] These can manifest in forms of 'power over', where we do something at the behest of an institution rather than ourselves. More insidiously, they can manifest through how we perceive the world around us. In other words, the process of social shaping is structured to form perceptions that benefit a few over the interests of all, including instrumental support from legal frameworks such as information privacy law regimes that play a 'constitutive role' in legitimating shaped constructions.[3]

A modulated society is consequently predicated on 'the economic and political strategies of powerful players' that flow everywhere through the 'values, priorities and assumptions' of technological design.[4] Citizenry and community in such a society is structured around profit-driven motives rather that selfhood-embarking provisions.[5] The collection of data about everything results in the dispersion and entanglement of boundary distinctions that once marked the identifiable parameters of liberal societies. If, as Cohen argues, modulation is the product of global informational capitalism,[6] then its consequences will lead to the vast relocation of structured power from the public to the private realm. Structured power is then able to manifest in the shaping of societal perceptions that 'private economic rights increasingly are understood as the

---

[2] Cohen, 'What Privacy Is For' 1913.
[3] Cohen, 'The Surveillance-Innovation Complex' 226.
[4] Cohen, 'The Networked Self in the Modulated Society' 71.
[5] Ibid.
[6] Ibid 74.

highest and best way of pursuing even important public purposes'.[7] The process of social shaping about the core provision of society and citizenry is itself shaped.

Such shaping also gives rise to significant questions about liberal democracy and its role of citizenship that flows to responsibilities and freedoms for individuals. Cohen argues strongly that liberal societies simply cannot function in the ever-attendant and pervading structures of surveillance that characterise informational capitalism.[8] The worthy liberal aspiration of the free-thinking and fully participatory individual cannot flourish in structures of modulated preference that have the effect of stymying difficult debate and replacing it with continually adjusted comfort zones.[9] It is the constant refinement, existence and justification of these comfort zones that blunts modulated citizenry and prevents it from asking more from its society than vapid forms of technologically driven convenience.[10]

Here resides the ultimate danger, the slide from unthinking, comfortable convenience into a modulated citizenry forming a predetermined democracy. But let us not leave the book's final considerations on this dystopian path, given Virginia Eubanks' helpful comments about positive solutions and indeed Julie Cohen's playful antidotes. Because the benefits of the smart world do not automatically have to result in modulated flows. Take, for example, Cohen's question

Will we be connected in ways that leave room to shape the contours of a 'new civility,' or will we be connected in ways dictated by the instrumental purposes of powerful global entities [?][11]

The tramlines of modulation's two-way flow are currently being laid. The uncritical and unthinking use of sensorisation that captures the obsolescence of the 'Internet of Things' does not assist us to work through such challenges. The 'things' are acts of power construction and delivery that seek to foster new modes and perspective about data collection. That said, modulation does not have to unfold at scale. The tramlines exist, but we can shape their ultimate terminus. But we need to do so actively and positively.

As individuals, this means being critical and questioning about what it means to have sensorised devices on our bodies, in our homes and in our environments. It also means thinking differently

---

7 Ibid 66.
8 Cohen, 'What Privacy Is For' 1912.
9 Ibid 1918.
10 Ibid.
11 Cohen, 'The Networked Self in the Modulated Society' 67.

about information privacy law and how its core protections could apply. Spaces for selfhood development are vital as an interruption of modulation. This should make us consider more deeply the conditions in which we flourish and modulated forms flow. The collected world analysis put forward in this book, even though largely descriptive in nature, assists us in viewing the smart world differently. It identifies the collected conditions that prepare modulated states. And by identifying the conditions more clearly, it becomes possible to develop potential individual, institutional and legal responses to modulation. Play, and playfulness, are important responses.

The play of everyday practice is the root of evolving subjectivity, and of creativity and innovation. It is how human beings pursue wellbeing and happiness in an imperfect and unpredictable world.[12]

Continuing in the vein of a positive conclusion, I will wrap up with two playful responses that highlight the importance of the collected world analysis and its role in interrupting modulation.

10.2   Institutional Tinkering

Chapter 9 suggested that information privacy law needs to encourage the articulation of privacy values as well as compliance with information privacy processes. Institutional culture is important in the development of technological innovation and design features that afford individual opportunities to interrupt modulation. Institutional tinkering thus becomes an important element of value articulation and design affordances. Tinkering refers to the playful posing of critical questions to inform institutional values about issues, such as privacy, that actively contribute to the institution's role in creating or obviating a modulated society. Tinkering disrupts institutional comfort zones. The resolution of these challenging questions and discussions then formulate into the value-based design of institutional data collection and analysis instruments. These thoughtful discussions, as Chapter 7 argued, are an institutionally focused cognitive interruption of modulation that poses an explicitly thoughtful set of values, inherent to the institution, which can be compared against the uncritical and unbidden logic of big data.

No commercial institution really knows what they are searching for in the collected world, which is why the notion of bioprospecting is an important one. Tinkering requires institutions to take a step back and

---

[12] Ibid 71.

ask, 'Is this what we should be doing?' rather than stating, 'This is what we have to do.' Again, as covered at many points in the book, this is a consequence of sensorisation, because behavioural patterns change, and the data collected says much more than even the institution wants to know about its customers.

Why would an institution go to such challenging lengths in eras of uncertainty, fear and change? Tinkering will provide a better alignment with existing and future customers about the institution's role as an information privacy co-creator of conditions that go beyond the emptiness of privacy policies as repetitions of self-interested compliance. A tinkered consideration of a privacy policy shifts from a legal risk management process of compliance, and even from one of modulation, to a thoughtful and critical discussion about the civic value of privacy that shapes involvement between the customer and the institution. The use of a privacy policy is no longer about procuring consent as a structure of prospective modulation and instead becomes a discussion about civic-value alignment in technological structures that neither the institution nor the customer truly understands.

The importance of institutional tinkering resides in its ability to identify and play in the interstitial gaps and spaces of legally and technologically constituted processes. Take, for example, my own research with Lizzie Coles-Kemp and Jodie Siganto.[13] In our qualitative interviews, we were interested in how a small number of Australian information security practitioners, mostly senior ones, constructed information security. The first question we asked at interviews, after covering educational and work histories, was 'What is your definition of information security?' The answer we got from most participants was typified as the 'old classic' – namely, 'confidentiality, integrity and availability'. There would then be a typical pregnant pause, after which the participant would furtively smile and say something like 'But that's not what my work is really about' and would go on to redefine information security considering their own experiences. The practitioners essentially engaged in forms of tinkering to try and work out what information security meant in their complex and conflict-ridden organisational contexts. The reframing of construct was a continuous act of tinkering driven by their own experience which had become part of their daily work, as benefiting their respective institutions.

---

[13] Mark Burdon, Jodie Siganto and Lizzie Coles-Kemp, 'The Regulatory Challenges of Australian Information Security Practice' (2016) 32 Computer Law & Security Review 623; Mark Burdon and Lizzie Coles-Kemp, 'The Significance of Securing as a Critical Component of Information Security: An Australian Narrative' (2019) 87 Computers & Security 1.

Tinkering for the practitioners was an attempt to secure a better alignment between compliance expectations and actualities. It is playful and thoughtful tinkering that leads to attempts at better alignment, because the players step out of the compliance-driven, institutional role to consider the actual effect on the organisation, including, most crucially, its customers. The participants also reported significant levels of conflict between different institutional groups. It is likely that some institutional groups and individuals will be more willing to tinker than others. However, if you recognise the value of tinkering as a broadly constituted institutional and civic good, then the critical tinkering of process play should be encouraged to better identify the issues that arise in the gaps between process boundaries.

Boundary considerations in the privacy sphere often emerge as 'creepy' forms of innovation that involve embedded tracking techniques to generate insight about customer behaviours. Creepiness arises because these tracking techniques generate sufficient uncertainty that they give rise to internal, ethical or moral concern. The shift from sole compliance of process to one of value confirmation will assist institutions to better articulate 'creepy' in line with the expectations of their customers.[14] It represents the movement from process to space in thinking critically about innovation. Tinkering activities should therefore receive the same institutional incentives as other readily established organisational components. However, that requires a clearly articulated value about the recognised purpose and good in tinkering and how it relates to broader understanding of the privacy values that represent the institution. Both would seem imperative elements of data collection strategies that potentially involve institutional collections of sensor data from the home.

10.3   Playful Weaving

It often seems that we as individuals can do little to change the structural forces that shape our lives. Globalisation, populism and modulation are all so vast that no one person could change them – or at least that's what we are led to believe. Cohen challenges us to think again, because the power of unpredictable play is a means for interrupting modulated power. Play reshapes the boundaries we make with institutions that are far more powerful than most of us. It is our unexpected

---

[14] Adam Thierer, 'The Pursuit of Privacy in a World Where Information Control Is Failing' (2013) 36 Harvard Journal of Law & Public Policy 409, 417–20 discussing creepiness as attempts to articulate privacy harm.

uses of technology that buck the segmented boxes which modulated forms of power would have us put in. The space for individual play is a boundary of the unexpected.

It is the unexpected that interrupts modulated forms of dictated governance, which become the basis of 'power over' activities. In sensor-based collections, the unexpected becomes harder, because the always-on collection capacities of environmental sensorisation have the scope to collect both the expected and the unexpected. In other words, if you constantly track behaviours, then the unexpected is something that eventually becomes expected and can therefore be predicted and prescripted. Something that is not normal can nonetheless be deemed normal over time if it unexpectedly emerges within a sufficiently identifiable pattern of behaviour. However, even if that is the case, unexpectedly playful acts create interstitial gaps and spaces where selfhood and reflection can occur. These are human spaces where we need to not be judged – by machine or by other humans.

A small anecdote helps explain. I have been lucky enough to have been assisted at various points of the book's journey by Tom Mackie. Tom told me a story that was relevant to the research on smart home insurance models. At that stage, he was a final-year law student and shared an apartment with three other male students. He was intrigued about the battery analysis capabilities of sensor devices such as Nest and Canary. When I asked why, he told me that the battery in his apartment's smoke alarm was recently running out of power. The alarm emitted a regular loud beep as a warning. Tom and his colleagues let the alarm beep for several days until one of them finally got sufficiently irritated by the noise and changed the battery.

Ignoring the battery warning beep for days is an unexpectedly playful act that only makes any kind of sense if you know that it unfolded in a student household – and probably a male-dominated student household at that. Were we to judge Tom and his colleagues, we could potentially use a range of equally judgemental adjectives such as lazy, risky, intransigent or even dangerous. Or we could use kinder terms such as playful, funny, youthful or experimental. There's no single human form of judgement to be ascribed, because that judging decision is part of how we ultimately understand our world and how that understanding shapes us.

Think now about how Tom and his colleagues would be assessed by a smart home business model of the type outlined in Chapter 4. The assessment would be defined by a model, which itself would presumably be based on the accumulation of similar data about smoke alarm battery uses or other smart home devices. Assessment is defined by institutional understanding of risk modelling rather than the kinder vagaries of human existence and its many perceptual forms. The assessment is also indelible where data about everything is not only sought but also retained. The several days of regular beeps could have a broader impact than its human form of equivalent judgement, which is ephemeral and generally passing.

Ignoring the battery beep is but one thread in a delicate weave of human understanding and selfhood. It is a weave that we all have a hand in making. We all weave the fabric of social life, and how much we play determines the quality of the weave. Spaces for play and tinkering are vital. It is in these cognitive breaks and interruptions that we can address some profound questions, such as who is providing the pattern for this weave? This is a key issue in the second decade of the twenty-first century. We weave away in our daily lives, but we are increasingly confused about the pattern we should follow. The level of disenchantment with the liberal society's political precepts and the complexity of the collected world is such that we could unintentionally be following a pattern of modulation – a pattern that is designed for the benefit of another – without realising it, and thus we don't ask whose pattern we are weaving.

That said, the quality of the weave is dependent upon each individual thread. Each thread contributes to something greater than the thread on its own. The same can be said of the complex societies that make up the collected world. The seamless flow of repetitive data pattern is the informational weave, clacking to and fro. The weave itself shapes understanding of the act of weaving and what we weave for. It is part of how we understand the weaving experience and how it, in turn, shapes our understanding about the pattern of weave we are creating.

Bibliography

Aldrich FK, 'Smart Homes: Past and Present' in Harper R (ed), *Inside the Smart Home* (Springer 2003).

Allen AL, 'Coercing Privacy' (1999) 40 *William and Mary Law Review* 723.

—'Privacy as Data Control: Conceptual, Practical and Moral Limits of the Paradigm' (2000) 32 *Connecticut Law Review* 861.

—*Unpopular Privacy: What Must We Hide?* (Oxford University Press 2012).

—'Synthesis and Satisfaction: How Philosophy Scholarship Matters' (2019) 20 *Theoretical Inquiries in Law* 343.

Andrejevic M, *iSpy: Power and Surveillance in the Interactive Era* (University of Kansas Press 2007).

Andrejevic M and Burdon M, 'Defining the Sensor Society' (2015) 16 *Television & New Media* 19.

Apthorpe N and others, 'Spying on the Smart Home: Privacy Attacks and Defenses on Encrypted IoT Traffic' <https://arxiv.org/pdf/1708.05044.pdf> accessed 20 June 2019.

Article 29 Data Protection Working Party, Opinion 4/2007 on the Concept of Personal Data (01248/07/EN WP 136, 2007).

—Opinion 13/2011 on Geolocation Services on Smart Mobile Devices (881/11/EN WP 185, 2011).

—Opinion 8/2014 on the Recent Developments on the Internet of Things (14/EN WP 223, 2014).

Austin L, 'Privacy and the Question of Technology' (2003) 22 *Law and Philosophy* 119.

—'Re-reading Westin' (2019) 20 *Theoretical Inquiries in Law* 1.

Australian Law Reform Commission, *For Your Information: Australian Privacy Law and Practice* (Law Reform Commission 2008).

Baillie L and Benyon D, 'Place and Technology in the Home' (2008) 17 *Computer Supported Cooperative Work* 227.

Balta-Ozkan N and others, 'The Development of Smart Homes Market in the UK' (2013) 60 *Energy* 361.

—'Social Barriers to the Adoption of Smart Homes' (2013) 63 *Energy Policy* 363.

Bannerman S, 'Relational Privacy and the Networked Governance of the Self' [2018] *Information, Communication & Society* 1.

Barlow J and Venables T, 'Smart Home, Dumb Suppliers? The Future of Smart Homes Markets' in Harper R (ed), *Inside the Smart Home* (Springer 2003).

299

Barnard-Wills D, Marinos L and Portesi S, *Threat Landscape and Good Practice Guide for Smart Home and Converged Media* (ENISA 2014).

Barocas S and Selbst A, 'Big Data's Disparate Impact' (2016) 104 *California Law Review* 671.

Bellia PL, 'Federalization in Information Privacy Law' (2009) 118 *Yale Law Journal* 868.

Benn SI, *A Theory of Freedom* (Cambridge University Press 1988).

Bennett CJ, *Regulating Privacy: Data Protection and Public Policy in Europe and the United States* (Cornell University Press 1992).

Bennett CJ and Raab CD, *The Governance of Privacy: Policy Instruments in Global Perspective* (MIT Press 2006).

Birchley G and others, 'Smart Homes, Private Homes? An Empirical Study of Technology Researchers' Perceptions of Ethical Issues in Developing Smart-Home Health Technologies' (2017) 18 *BMC Medical Ethics* 1.

Birnhack M, 'A Process-Based Approach to Informational Privacy and the Case of Big Medical Data' (2019) 20 *Theoretical Inquiries in Law* 257.

Bloustein EJ, 'Privacy as an Aspect of Human Dignity: An Answer to Dean Prosser' (1964) 39 *New York University Law Review* 962.

Booth S and others, *What Are 'Personal Data'? A Study Conducted for the UK Information Commissioner* (Information Commissioner Office 2004).

Braun A and Scheriber F, *The Current InsurTech Landscape: Business Models and Disruptive Potential* (University of St. Gallen 2016).

Brich J and others, 'Exploring End User Programming Needs in Home Automation' (2017) 24 *ACM Transactions on Computer-Human Interaction* 1.

Brynjolfsson E and McAfee A, *The Second Machine Age: Work, Progress, and Prosperity in a Time of Brilliant Technologies* (W. W. Norton & Company 2014).

Bryson S and others, 'Visually Exploring Gigabyte Data Sets in Real Time' (1999) 42 *Communications of the ACM* 82.

Bugeja J, Jacobsson A and Davidsson P, 'On Privacy and Security Challenges in Smart Connected Homes' (European Intelligence and Security Informatics Conference, Upsalla, Sweden, August 2016).

Burdon M, 'Contextualizing the Tensions and Weaknesses of Data Breach Notification and Information Privacy Law' (2010) 27 *Santa Clara Computer and High Technology Law Journal* 63.

—'Privacy Invasive Geo-mashups: Privacy 2.0 and the Limits of First Generation Information Privacy Laws' [2010] *University of Illinois Journal of Law, Technology and Policy* 1.

Burdon M and Andrejevic M, 'Big Data in the Sensor Society' in Ekbia H, Sugimoto C and Mattiolli M (eds), *Big Data Is Not a Monolith* (MIT Press 2016).

Burdon M and Coles-Kemp L, 'The Significance of Securing as a Critical Component of Information Security: An Australian Narrative' (2019) 87 *Computers & Security* 1.

Burdon M and Harpur P, 'Re-conceptualising Privacy and Discrimination in an Age of Talent Analytics' (2014) 37 *University of New South Wales Law Journal* 679.

Burdon M and McKillop A, 'The Google Street View Wi-Fi Scandal and Its Repercussions for Privacy Regulation' (2013) 39 *Monash University Law Review* 702.

Burdon M, Siganto J and Coles-Kemp L, 'The Regulatory Challenges of Australian Information Security Practice' (2016) 32 *Computer Law & Security Review* 623.

Burdon M and Telford P, 'The Conceptual Basis of Personal Information in Australian Privacy Law' (2010) 17 *Murdoch Elaw Journal* 1.

Bygrave LA, *Data Protection Law: Approaching Its Rationale, Logic and Limits* (Information Privacy, Kluwer Law International 2002).

—*Data Privacy Law: An International Perspective* (Oxford University Press 2014).

Canaan M, Lucker J and Spector B, *Opting In: Using IoT Connectivity to Drive Differentiation* (Deloitte University Press 2016).

Cappiello A, *Technology and the Insurance Industry Re-configuring the Competitive Landscape* (Springer International Publishing 2018).

Cate FH, 'The EU Data Protection Directive, Information Privacy, and the Public Interest' [1994] *Iowa Law Review* 431.

Cate FH and Mayer-Schönberger V, 'Notice and Consent in a World of Big Data' (2013) 3 *International Data Privacy Law* 67.

CB Insights, *How Major Insurers Are Teaming Up with Internet of Things Companies in One Infographic* (CB Insights 2015) <www.cbinsights.com/research/insurance-tech-iot-partnerships/>.

Cebulsky M and others, 'The Digital Insurance: Facing Customer Expectation in a Rapidly Changing World' in Linnhoff-Popien C, Schneider R and Zaddach M (eds), *Digital Marketplaces Unleashed* (Springer 2017).

Chen Z and others, 'Unobtrusive Sleep Monitoring Using Smartphones' (Proceedings of the 2013 7th International Conference on Pervasive Computing Technologies for Healthcare and Workshops, PervasiveHealth, Venice, May 2013).

Choudhury T and Pentland S, 'Sensing and Modeling Human Networks Using the Sociometer' (Seventh IEEE International Symposium on Wearable Computers, ISWC 2003, October 2003, White Plains, New York).

Clarke R and Greenleaf G, 'Dataveillance Regulation: A Research Framework' (2017) 25 *Journal of Law, Information and Science* 104.

Cocchia A, 'Smart and Digital City: A Systematic Literature Review' in Dameri PR and Rosenthal-Sabroux C (eds), *Smart City: How to Create Public and Economic Value with High Technology in Urban Space* (Springer International Publishing 2014).

Coffman K and Odlyzko A, 'The Size and Growth of the Internet' (1998) 3 *First Monday*.

Cognizant, *Next-Generation Insurance: Tapping into the Intelligence of Smart Homes* (2015) <www.cognizant.com/InsightsWhitepapers/next-generation-insurance-tapping-into-the-intelligence-of-smart-homes-codex1411.pdf>.

Cohen JE, 'Examined Lives: Informational Privacy and the Subject as Object' (2000) 52 *Stanford Law Review* 1373.

—*Configuring the Networked Self: Law, Code, and the Play of Everyday Practice* (Yale University Press 2012).

—'What Privacy Is For' (2013) 126 *Harvard Law Review* 1904.

—'Between Truth and Power' in Hildebrandt M and van den Berg B (eds), *Freedom and Property of Information: The Philosophy of Law Meets the Philosophy of Technology* (Routledge 2014).

—'The Networked Self in the Modulated Society' in de Been W, Arora P and Hildebrandt M (eds), *Crossroads in New Media, Identity and Law: The Shape of Diversity to Come* (Palgrave Macmillan UK 2015).

—'Affording Fundamental Rights: A Provocation Inspired by Mireille Hildebrandt' (2016) 4 *Critical Analysis of Law* 78.

—'The Surveillance-Innovation Complex: The Irony of the Participatory Turn' in Barney D and others (eds), *The Participatory Condition in the Digital Age* (Minnesota University Press 2016).

—'The Biopolitical Public Domain: The Legal Construction of the Surveillance Economy' (2018) 31 *Philosophy & Technology* 213.

—'Review of Zuboff, Shoshana. 2019. The Age of Surveillance Capitalism: The Fight for a Human Future at the New Frontier of Power' (2019) 17 *Surveillance & Society* 240.

—'Turning Privacy Inside Out' (2019) 20 *Theoretical Inquiries in Law* 1.

Cortis D and others, 'InsurTech' in Lynn T and others (eds), *Disrupting Finance: FinTech and Strategy in the 21st Century* (Springer International Publishing 2019).

Costa L, *Virtuality and Capabilities in a World of Ambient Intelligence: New Challenges to Privacy and Data Protection* (Springer 2016).

Cox E, *Retail Analytics: The Secret Weapon* (Wiley 2012).

Crawford K and Schultz J, 'Big Data and Due Process: Toward a Framework to Redress Predictive Privacy Harms' (2014) 55 *Boston College Law Review* 93.

Davenport N, 'Smart Washers May Clean Your Clothes, but Hacks Can Clean Out Your Privacy, and Underdeveloped Regulations Could Leave You Hanging on a Line' (2016) 32 *The John Marshall Journal of Information Technology & Privacy Law*.

Davenport TH, *The New World of Business Analytics* (International Institute of Analytics 2010).

—'Analytics 3.0' (2013) 91 *Harvard Business Review* 64.

Davenport TH, Barth P and Bean R, 'How Big Data Is Different' (2012) 54 *MIT Sloan Management Review* 43.

Davenport TH and Dyche J, *Big Data in Big Companies* (International Institute for Analytics 2013).

Davenport TH, Harris J and Morrison R, *Analytics at Work* (Harvard Business Press 2010).

Davenport TH, Harris J and Shapiro J, 'Competing on Talent Analytics' (2010) 88 *Harvard Business Review* 52.

Davenport TH and Patil DJ, 'Data Scientist: The Sexiest Job of the 21st Century' (2012) 90 *Harvard Business Review* 70.

Davidoff S and others, 'Principles of Smart Home Control' in Dourish P and Friday A (eds), *UbiComp 2006: Ubiquitous Computing* (Springer Berlin Heidelberg 2006).

DeCew JW, 'The Conceptual Coherence of Privacy as Developed in Law' in Cudd AE and Navin MC (eds), *Core Concepts and Contemporary Issues in Privacy* (Springer 2018).

Demiris G and Hensel B, 'Technologies for an Aging Society: A Systematic Review of "Smart Home" Applications' [2008] *IMIA Yearbook of Medical Informatics* 33.

—'Smart Homes for Patients at the End of Life' (2009) 23 *Journal of Housing For the Elderly* 106.

Denning PJ, 'Saving All the Bits' (1990) 78 *American Scientist* 402.

Dickinson P and others, 'Indoor Positioning of Shoppers Using a Network of Bluetooth Low Energy Beacons' (2016 International Conference on Indoor Positioning and Indoor Navigation (IPIN), October 2016, Alcalá de Henares, Spain).

Dinev T and others, 'Information Privacy and Correlates: An Empirical Attempt to Bridge and Distinguish Privacy-Related Concepts' (2013) 22 *European Journal of Information Systems* 295.

Dirks S and Keeling M, *A Vision of Smarter Cities: How Cities Can Lead the Way into a Prosperous and Sustainable Future* (IBM Global Business Services 2009).

Dowding K, 'Three-Dimensional Power: A Discussion of Steven Lukes' 'Power: A Radical View' (2006) 4 *Political Studies Review* 136.

Draper NA and Turow J, 'The Corporate Cultivation of Digital Resignation' [2019] *New Media & Society*.

Dyrberg TB, *The Circular Structure of Power: Politics, Identity, Community* (Verso 1997).

Edwards L, 'Privacy, Security and Data Protection in Smart Cities: A Critical EU Law Perspective' (2016) 2 *European Data Protection Law Review* 28.

Edwards L and Veale M, 'Slave to the Algorithm? Why a 'Right to Explanation' Is Probably Not the Remedy You Are Looking For' (2018) 16 *Duke Law and Technology Review* 18.

Ehrenreich R, 'Privacy and Power' (2001) 89 *Georgetown Law Journal* 2047.

Eling M and Lehmann M, 'The Impact of Digitalization on the Insurance Value Chain and the Insurability of Risks' [2018] 43 *The Geneva Papers on Risk and Insurance – Issues and Practice* 359–396.

Eubanks V, *Digital Dead End: Fighting for Social Justice in the Information Age* (MIT Press 2011).

— *Automating Inequality: How High-Tech Tools Profile, Police, and Punish the Poor* (1st edn, St. Martin's Press 2018).

European Data Protection Supervisor, *Privacy and Competitiveness in the Age of Big Data: The Interplay between Data Protection, Competition Law and Consumer Protection in the Digital Economy* (European Data Protection Supervisor 2014).

Ferguson AG, 'The Internet of Things and the Fourth Amendment of Effects' (2016) 104 *California Law Review* 805.

Forgó N, Hänold S and Schütze B, 'The Principle of Purpose Limitation and Big Data' in Corrales M, Fenwick M and Forgó N (eds), *New Technology, Big Data and the Law* (Springer 2017).

Fried C, 'Privacy' (1968) 77 *Yale Law Journal* 475.

Gabrys J, 'Citizen Sensing, Air Pollution and Fracking: From "Caring About Your Air" to Speculative Practices of Evidencing Harm' (2017) 65 *The Sociological Review* 172.

Gandy OH and Nemorin S, 'Toward a Political Economy of Nudge: Smart City Variations' [2018] *Information, Communication & Society* 1.

Gavison R, 'Privacy and the Limits of Law' (1980) 89 *Yale Law Journal* 421.

Gellman R, 'Does Privacy Law Work?' in Agre P and Rotenberg M (eds), *Technology and Privacy: The New Landscape* (Information Privacy, MIT Press 1997).

Geneiatakis D and others, 'Security and Privacy Issues for an IoT Based Smart Home' (40th International Convention on Information and Communication Technology, Electronics and Microelectronics (MIPRO), Opatijia, Croatia, May 2017).

Goldsmith S and Crawford S, *The Responsive City* (Jossey-Bass 2014).

Gonzâlez FusterG, *The Emergence of Personal Data Protection as a Fundamental Right of the EU* (Springer 2014).

Gormley K, 'One Hundred Years of Privacy' [1992] *Wisconsin Law Review* 1335.

Graham-Rowe D, 'A Smart Phone That Knows You're Angry' *MIT Technology Review* <www.technologyreview.com/s/426560/a-smart-phone-that-knows-youre-angry/> accessed 2 May 2019.

Greenleaf G, *Balancing Globalisation's Benefits and Commitments: Accession to Data Protection Convention 108 by Countries Outside Europe* (UNSW Law Research Paper No 16-52, 2016).

Greenleaf GW, 'Privacy in Australia' in Greenleaf GW and Rule JB (eds), *Global Privacy Protection: The First Generation* (Edward Elgar Publishing Limited 2008).

Gross H, 'The Concept of Privacy' (1967) 42 *New York University Law Review* 34.

Gurses S and van Hoboken J, 'Privacy after the Agile Turn' <https://osf.io/preprints/socarxiv/9gy73/> accessed 20 June 2019.

Harper R, *Inside the Smart Home* (Springer 2003).

—'From Smart Home to Connected Home' in Harper R (ed), *The Connected Home: The Future of Domestic Life* (Springer 2011).

Hartzog W, 'Opinions · The Case against Idealising Control' (2018) 4 *European Data Protection Law Review* 423.

Hartzog W and Solove DJ, 'The Scope and Potential of FTC Data Protection' (2015) 83 *George Washington Law Review* 2230.

Hildebrandt M, 'Law as Information in the Era of Data-Driven Agency' (2016) 79 *The Modern Law Review* 1.

Hiller JS and Blanke JM, 'Smart Cities, Big Data, and the Resilience of Privacy' (2017) 68 *Hastings Law Journal* 309.

Hollands RG, 'Critical Interventions into the Corporate Smart City' (2015) 8 *Cambridge Journal of Regions, Economy and Society* 61.

Hoogendoorn M and Funk B, *Machine Learning for the Quantified Self on the Art of Learning from Sensory Data* (Springer 2018).

Huijsing JH, 'Smart Sensor Systems: Why? Where? How?' in Meijer G (ed), *Smart Sensor Systems* (John Wiley & Sons 2008).

Information Commissioner's Office (UK), *Guide to the General Data Protection Regulation (GDPR)* (2018) <https://ico.org.uk/for-organisations/guide-to-data-protection/guide-to-the-general-data-protection-regulation-gdpr/>.

Inness JC, *Privacy, Intimacy, and Isolation* (Oxford University Press 1992).

Jacobsson A, Boldt M and Carlsson B, 'A Risk Analysis of a Smart Home Automation System' (2016) 56 *Future Generation Computer Systems* 719.

Jacobsson A and Davidsson P, 'Towards a Model of Privacy and Security for Smart Homes' (IEEE 2nd World Forum on Internet of Things (WF-IoT), Milan, December 2015).

Jones ML, 'Privacy without Screens & the Internet of Other People's Things' (2015) 51 *Idaho Law Review* 639.

Kamiran F and Calders T, 'Data Preprocessing Techniques for Classification without Discrimination' (2012) 33 *Knowledge and Information Systems* 1.

Kang J, 'Information Privacy in Cyberspace Transactions' (1998) 50 *Stanford Law Review* 1193.

Katyal S, 'Privacy vs. Piracy' (2004) 7 *Yale Journal of Law & Technology* 222.

Keats Citron D, 'Technological Due Process' (2007) 85 *Washington University Law Review* 1249.

Kellmereit D and Obodovski D, *The Silent Intelligence: The Internet of Things* (DND Ventures LLC 2013).

Kerr I and Earle J, 'Prediction, Preemption, Presumption: How Big Data Threatens Big Picture Privacy' 66 *Stanford Law Review Online* 65.

Kirby M, 'Twenty-five Years of Evolving Information Privacy Law: Where Have We Come From and Where Are We Going?' (2003) 21 *Prometheus* 467.

Kitchin R, 'The Real-Time City? Big Data and Smart Urbanism' (2014) 79 *GeoJournal* 1.

Knopf GK and Bassi AS, 'Introduction to Biosensors and Bioelectronics' in Knopf GK and Bassi AS (eds), *Smart Biosensor Technology* (2nd edn, CRC Press 2019).

Koops B-J, 'The Trouble with European Data Protection Law' (2014) 4 *International Data Privacy Law* 250.

Koops B-J and others, 'A Typology of Privacy' (2017) 38 *University of Pennsylvania Journal of International Law* 483.

Kranz M, *Building the Internet of Things: Implement New Business Models, Disrupt Competitors, and Transform Your Industry* (Wiley 2017).

Lane M, 'Location-Based Analytics Yield Customer and Inventory Insights' 4 *Journal of Retail Analytics* 21.

Lanzing M, 'The Transparent Self' (2016) 18 *Ethics and Information Technology* 9.

Larsen NM, Sigurdsson V and Breivik J, 'The Use of Observational Technology to Study In-Store Behavior: Consumer Choice, Video Surveillance, and Retail Analytics' (2017) 40 *The Behavior Analyst* 343.

Latif MA and others, 'User Privacy Framework for Web-of-Objects Based Smart Home Services' (2015) 9 *International Journal of Smart Home* 61.

Lessig L, 'Privacy as Property' (2002) 69 *Social Research* 247.

Lewis S, 'Insurtech: An Industry Ripe for Disruption' (2017) 1 *The Georgetown Law Technology Review* 491.

Lewis SCR, 'Energy in the Smart Home' in Harper R (ed), *The Connected Home: The Future of Domestic Life* (Springer 2011).

Liddle J and others, 'Measuring the Lifespace of People with Parkinson's Disease Using Smartphones: Proof of Principle' (2014) 16 *Journal of Medical Internet Research* e13.

Likamwa R and others, 'MoodScope: Building a Mood Sensor from Smartphone Usage Patterns' (Proceeding of the 11th Annual International Conference on Mobile Systems, Applications, and Services, MobiSys '13, Taipei, June 2013).

Lin H and Bergmann N, 'IoT Privacy and Security Challenges for Smart Home Environments' (2016) 7 *Information* 44.

Lindsay D, 'An Exploration of the Conceptual Basis of Privacy and the Implications for the Future of Australian Privacy Law' (2005) 29 *Melbourne University Law Review* 131.

Linnhoff-Popien C, Schneider R and Zaddach M, *Digital Marketplaces Unleashed* (Springer 2017).

Litman J, 'Information Privacy/Information Property' (2000) 52 *Stanford Law Review* 1283.

Löffler M and others, *Insurers Need to Plug into the Internet of Things: Or Risk Falling Behind* (McKinsey & Company 2016).

Luhn HP, 'A Business Intelligence System' (1958) 2 *IBM Journal of Research and Development* 314.

Lukes S, *Power: A Radical View* (Palgrave Macmillan 1974).

—*Power: A Radical View* (2nd edn, Palgrave Macmillan 2005).

Lupton D, 'Feeling Your Data: Touch and Making Sense of Personal Digital Data' (2017) 19 *New Media & Society* 1599.

Lyman P and Varian H, 'How Much Information?' (2000) 6 *JEP: The Journal of Electronic Publishing*.

Lynskey O, 'Aligning Data Protection Rights with Competition Law Remedies? The GDPR Right to Data Portability' (2017) 42 *European Law Review* 793.

MacCarthy M, 'In Defense of Big Data Analytics' in Selinger E, Polonetsky J and Tene O (eds), *The Cambridge Handbook of Consumer Privacy* (Cambridge University Press 2018).

Mai H, 'Preface: Fin- & Insuretech' in Linnhoff-Popien C, Schneider R and Zaddach M (eds), *Digital Marketplaces Unleashed* (Springer 2017).

Maisel L and CokinsG, *Predictive Business Analytics: Forward-Looking Capabilities to Improve Business Performance* (John Wiley & Sons 2014).

Margulis ST, 'On the Status and Contribution of Westin's and Altman's Theories of Privacy' (2003) 59 *Journal of Social Issues* 411.

McFall L and Moor L, 'Who, or What, Is Insurtech Personalizing? Persons, Prices and the Historical Classifications of Risk' (2018) 19 *Distinktion: Journal of Social Theory* 193.

Mendes TDP and others, 'Smart Home Communication Technologies and Applications: Wireless Protocol Assessment for Home Area Network Resources' (2015) 8 *Energies* 7279.

Miller AR, 'Personal Privacy in the Computer Age: The Challenge of a New Technology in an Information-Oriented Society' (1968) 67 *Michigan Law Review* 1091.

Mittelstadt B, 'Designing the Health-Related Internet of Things: Ethical Principles and Guidelines' (2017) 8 *Information* 77.

Montgomery AR, 'Just What the Doctor Ordered: Protecting Privacy without Impeding Development of Digital Pills' (2016) 19 *Vanderbilt Journal of Entertainment and Technology Law* 147.

Moor JH, 'Towards a Theory of Privacy in the Information Age' (1997) 27 *Computers and Society* 27.

Moore A, 'Privacy, Speech and the Law' (2013) 22 *Journal of Information Ethics* 21.

Moore AD, 'Intangible Property: Privacy, Power and Control' in Moore AD (ed), *Information Ethics: Privacy, Property, and Power* (University of Washington Press 2005).

Mulligan DK, Koopman C and Doty N, 'Privacy Is an Essentially Contested Concept: A Multi-Dimensional Analytic for Mapping Privacy' (2016) 374 *Philosophical Transactions of the Royal Society A* 1.

Mundie C, 'Privacy Pragmatism' (2014) 93 *Foreign Affairs* 28.

Munich RE and Hartford Steam Boiler, *Connected Home Technologies* (2016) <www.munichre.com/site/hsb/get/documents_E-1898694672/hsb/assets.hsb.group/Documents/Knowledge-Center/Equipment-Care-for-Homeowners/connected-home-technologies.pdf>.

Murakami Wood D and Mackinnon D, 'Partial Platforms and Oligoptic Surveillance in the Smart City' (2019) 17 *Surveillance & Society* 176.

Murphy RS, 'Property Rights in Personal Information: An Economic Defense of Privacy' (1996) 84 *Georgetown Law Journal* 2381.

Naylor M, *Insurance Transformed* (Springer 2017).

Neff G and Nafus D, *Self-Tracking* (The MIT Press 2016).

Newell B, 'Privacy and Surveillance in the Streets' in Newell BC, Timan T and Koops B-J (eds), *Surveillance, Privacy and Public Space* (Routledge 2019).

Nicoletti B, *The Future of Fintech: Integrating Finance and Technology in Financial Services* (Palgrave Macmillan 2017).

Nissenbaum H, 'Protecting Privacy in an Information Age: The Problem of Privacy in Public' (1998) 17 *Law and Philosophy* 559.

—'Privacy as Contextual Integrity' (2004) 79 *Washington Law Review* 119.

—*Privacy in Context: Technology, Policy, and the Integrity of Social Life* (Stanford Law Books 2010).

—'Must Privacy Give Way to Use Regulation?' 2015–16 Faculty Seminar: Democracy, Citizenship, and Constitutionalism <www.sas.upenn.edu/andrea-mitchell-center/sites/www.sas.upenn.edu.dcc/files/Nissenbaum-UPenn-Democracy.pdf> accessed 14 May 2019.

Nyarku M and others, 'Mobile Phones as Monitors of Personal Exposure to Air Pollution: Is This the Future?' (2018) 13 *PLoS One* e0193150.

OECD, *Guidelines on the Protection of Privacy and Transborder Flows of Personal Data* (OECD 1980).

—*Smart Sensor Networks: Technologies and Applications for Green Growth* (OECD Digital Economy Papers 2009).

—*Machine-to-Machine Communications: Connecting Billions of Devices* (OECD Digital Economy Papers 2012).

—*The OECD Privacy Framework* (OECD Publishing 2013).

—*Emerging Issues: The Internet of Things* (Digital Economy Outlook 2015).

—*Technology and Innovation in the Insurance Sector* (OECD 2017).

Office of the Australian Information Commissioner, *Australian Privacy Principles Guidelines* (OAIC 2014).
—*What Is Personal Information?* (OAIC 2017).
Ohm P, 'Broken Promises of Privacy: Responding to the Surprising Failure of Anonymization' 57 *UCLA Law Review* 1701.
Pasquale F, 'Restoring Transparency to Automated Authority' (2011) 9 *Journal on Telecommunications and High Technology Law* 235.
Paterson M and McDonagh M, 'Data Protection in an Era of Big Data: The Challenges Posed by Big Personal Data' (2018) 44 *Monash University Law Review* 1.
Peek ME, 'Information Privacy and Corporate Power: Towards a Re-Imagination of Information Privacy Law' (2006) 37 *Seton Hall Law Review* 127.
Peppet SR, 'Regulating the Internet of Things: First Steps Toward Managing Discrimination, Privacy, Security, and Consent' (2014) 93 *Texas Law Review* 85.
Perera C and others, 'Sensing as a Service Model for Smart Cities Supported by Internet of Things' (2013) 25 *Transactions on Emerging Telecommunications Technologies* 81.
Plachkinova M, Vo A and Alluhaidan A, 'Emerging Trends in Smart Home Security, Privacy, and Digital Forensics' AMCIS (2016) <https://aisel.aisnet.org/cgi/viewcontent.cgi?article=1434&context=amcis2016>.
Porter M and Heppelmann J, 'Connected Products Are Transforming Competition' (2014) Harvard Business Review 1.
Posner RA, 'The Right of Privacy' (1978) 12 *Georgia Law Review* 393.
Post RC, 'Three Concepts of Privacy' (2001) 89 *Georgetown Law Journal* 2087.
Poullet Y, 'Is the General Data Protection Regulation the Solution?' (2018) 34 *Computer Law & Security Review* 773.
President's Council of Advisors on Science and Technology, *Big Data and Privacy: A Technological Perspective* (White House Office of Science & Technology Policy 2014).
Price Waterhouse Coopers, *Insurance's New Normal: Driving Innovation with InsurTech* (2016) <www.pwc.com/gx/en/industries/financial-services/fintech-survey/report/insurance.html>.
—*InsurTech: A Force for Good* (2016) <www.pwc.co.uk/industries/financial-services/insurtech.html>.
—*Opportunities Await: How InsurTech Is Reshaping Insurance* (Global FinTech Survey, June 2016).
Prins C, 'Property and Privacy: European Perspectives and the Commodification of Our Identity' in Guibault LMCR and Hugenholtz PB (eds), *The Future of the Public Domain* (Information Privacy, Kluwer Law International 2006).
Purtova N, 'The Illusion of Personal Data as No One's Property' (2015) 7 *Law, Innovation and Technology* 83.
—'The Law of Everything: Broad Concept of Personal Data and Future of EU Data Protection Law' (2018) 10 *Law, Innovation and Technology* 40.
Raul AC, *Privacy and the Digital State: Balancing Public Information and Personal Privacy* (Kluwer 2002).
Regan PM, *Legislating Privacy: Technology, Social Values, and Public Policy* (University of North Carolina Press 1995).

—'Privacy and the Common Good: Revisited' in Roessler B and Mokrosinska D (eds), *Social Dimensions of Privacy* (Cambridge University Press 2015).

Reidenberg J and others, 'Disagreeable Privacy Policies: Mismatches between Meaning and Users' Understanding' (2015) 30 *Berkeley Technology Law Journal* 39.

Reidenberg JR, 'Privacy in the Information Economy: A Fortress or Frontier for Individual Rights?' (1992) 44 *Federal Communications Law Journal* 195.

Report of the Secretary's Advisory Committee on Automated Personal Data Systems, *Records, Computers and the Rights of Citizens* (United States Government Printing Office 1973).

Riemer K and others, *The Fintech Advantage: Harnessing Digital Technology, Keeping the Customer in Focus* (Capgemini & The University of Sydney Business School 2017).

Risteska Stojkoska BL and Trivodaliev KV, 'A Review of Internet of Things for Smart Home: Challenges and Solutions' (2016) 140 *Journal of Cleaner Production* 1454.

Roberts A, 'Why Privacy and Domination' (2018) 4 *European Data Protection Law Review* 5.

Rosen J, *The Unwanted Gaze: The Destruction of Privacy in America* (Random House 2000).

Rotenberg M, 'Fair Information Practices and the Architecture of Privacy' *Stanford Technology Law Review* <http://stlr.stanford.edu/STLR/Articles/ 01_STLR_1> accessed 20 June 2019.

Rubenfeld J, 'The Right of Privacy and the Right to Be Treated as an Object' (2001) 89 *Georgetown Law Journal* 2099.

Rubinstein I, 'Privacy Localism' (2018) 93 *Washington Law Review* 1961.

Rubinstein IS, Lee RD and Schwartz PM, 'Data Mining and Internet Profiling: Emerging Regulatory and Technological Approaches' (2008) 75 *The University of Chicago Law Review* 261.

Rule JB, *Privacy in Peril* (Oxford University Press 2007).

Saad Al-Sumaiti A, Ahmed MH and Salama MMA, 'Smart Home Activities: A Literature Review' (2014) 42 *Electric Power Components and Systems* 294.

Saxena RN and Srinivasan A, *Business Analytics: A Practitioner's Guide* (Springer 2013).

Schauer L, 'Analyzing the Digital Society by Tracking Mobile Customer Devices' in Linnhoff-Popien C, Schneider R and Zaddach M (eds), *Digital Marketplaces Unleashed* (Springer 2017).

Schermer B, 'The Limits of Privacy in Automated Profiling and Data Mining' (2011) 27 *Computer Law & Security Review* 45.

Schoeman FD, *Privacy and Social Freedom* (Information Privacy, Cambridge University Press 1992).

Schwartz PM, 'Privacy and Democracy in Cyberspace' (1999) 52 *Vanderbilt Law Review* 1609.

—'Internet Privacy and the State' (2000) 32 *Connecticut Law Review* 815.

—'Preemption and Privacy' (2009) 118 *Yale Law Journal* 902.

Schwartz PM and Janger EJ, 'Notification of Data Security Breaches' (2007) 105 *Michigan Law Review* 913.

Schwartz PM and Peifer K-N, 'Transatlantic Data Privacy Law' (2017) 106 *Georgetown Law Journal* 115.

Schwartz PM and Solove DJ, 'Reconciling Personal Information in the United States and European Union' (2014) 102 *California Law Review* 877.

Scoble R and Israel S, *Age of Context: Mobile, Sensors, Data and the Future of Privacy* (Patrick Brewster Press 2014).

Sedenberg E, Wong R and Chuang J, 'A Window into the Soul' in Newell BC, Timan T and Koops B-J (eds), *Surveillance, Privacy and Public Space* (Routledge 2019).

Shackelford SJ and others, 'When Toasters Attack: A Polycentric Approach to Enhancing the Security of Things' (2017) 2017 *University of Illinois Law Review* 415.

Silk M, 'Beacons Are Not the New Black' (2015) 4 *Journal of Retail Analytics* 6.

Simitis S, 'Reviewing Privacy in an Information Society' (1987) 135 *University of Pennsylvania Law Review* 707.

Sinclair B, *IoT Inc: How Your Company Can Use the Internet of Things to Win in the Outcome Economy* (McGraw-Hill Education 2017).

Sivaraman V, Gharakheili HH and Fernandes C, *Inside Job: Security and Privacy Threats for Smart-home IoT Devices* (University of New South Wales and Australian Communications Consumer Action Network (ACCAN) 2017) <http://accan.org.au/files/Grants/UNSW-ACCAN_InsideJob_web.pdf>.

Smith GJD, 'Data Doxa: The Affective Consequences of Data Practices' (2018) 5 *Big Data & Society* 1.

Solaimani S, Keijzer-Broers W and Bouwman H, 'What We Do – and Don't – Know about the Smart Home: An Analysis of the Smart Home Literature' (2015) 24 *Indoor and Built Environment* 370.

Solove DJ, 'Privacy and Power: Computer Databases and Metaphors for Information Privacy' (2001) 53 *Stanford Law Review* 1393.

—'Conceptualizing Privacy' (2002) 90 *California Law Review* 1087.

—'"I've Got Nothing to Hide" and Other Misunderstandings of Privacy' (2008) 44 *San Diego Law Review* 745.

—*Understanding Privacy* (Harvard University Press 2008).

—'Privacy Self-Management and the Consent Dilemma' (2013) 126 *Harvard Law Review* 1880.

Solove DJ and Hartzog W, 'The FTC and the New Common Law of Privacy' (2014) 114 *Columbia Law Review* 583.

Solove DJ and Schwartz PM, 'The PII Problem: Privacy and a New Concept of Personally Identifiable Information' (2011) 86 *New York University Law Review* 1814.

Sprague R, 'Welcome to the Machine: Privacy and Workplace Implications of Predictive Analytics' (2015) 21 *Richmond Journal of Law & Technology* 1.

Steeves V, 'Theorizing Privacy in a Liberal Democracy: Canadian Jurisprudence, Anti-Terrorism, and Social Memory after 9/11' (2019) 20 *Theoretical Inquiries in Law* 323.

Strain J, 'Households as Morally Ordered Communities: Explorations in the Dynamics of Domestic Life' in Harper R (ed), *Inside the Smart Home* (Springer 2003).

Strandburg KJ, 'Monitoring, Datafication and Consent: Legal Approaches to Privacy in the Big Data Context' in Lane J and others (eds), *Privacy, Big Data and the Public Good: Frameworks for Engagement* (Cambridge University Press 2014).

Susser D, 'Notice after Notice-and-Consent: Why Privacy Disclosures Are Valuable Even If Consent Frameworks Aren't' [2019] 9 *Journal of Information Policy* 37.

Swanson B and Gilder G, *Estimating the Exaflood: The Impact of Video and Rich Media on the Internet* (Discovery Institute 2008).

Takayama L and others, 'Making Technology Homey: Finding Sources of Satisfaction and Meaning in Home Automation' (UbiComp'12 – Proceedings of the 2012 ACM Conference on Ubiquitous Computing, Pittsburgh, September 2012).

Tavani HT, 'Philosophical Theories of Privacy: Implications for an Adequate Online Privacy Policy' (2007) 38 *Metaphilosophy* 1.

Tene O and Polonetsky J, 'Big Data for All: Privacy and User Control in the Age of Analytics' (2013) 11 *Northwestern Journal of Technology and Intellectual Property* 239.

—'Big Data for All: Privacy and User Control in the Age of Analytics' (2013) 11 *Northwestern Journal of Technology and Intellectual Property* 240.

Theoharidou M, Tsalis N and Gritzalis D, 'Smart Home Solutions: Privacy Issues' in van Hoof J, Demiris G and Wouters EJM (eds), *Handbook of Smart Homes, Health Care and Well-Being* (Springer 2014).

Thierer A, 'The Pursuit of Privacy in a World Where Information Control Is Failing' (2013) 36 *Harvard Journal of Law & Public Policy* 409.

Thorne C and Griffiths C, 'Smart, Smarter, Smartest: Redefining Our Cities' in Dameri PR and Rosenthal-Sabroux C (eds), *Smart City: How to Create Public and Economic Value with High Technology in Urban Space* (Springer International Publishing 2014).

Tolmie P and Crabtree A, 'The Practical Politics of Sharing Personal Data' (2018) 22 *Personal and Ubiquitous Computing* 293.

Tolmie P and others, 'Towards the Unremarkable Computer: Making Technology at Home in Domestic Routines' in Harper R (ed), *Inside the Smart Home* (Springer 2003).

Toschi GM, Campos LB and Cugnasca CE, 'Home Automation Networks: A Survey' (2017) 50 *Computer Standards & Interfaces* 42.

Tovino SA, 'The HIPAA Privacy Rule and the EU GDPR: Illustrative Comparisons' (2017) 47 *Seton Hall Law Review* 993.

Townsend A and Arthur J, *Smart Cities: Big Data, Civic Hackers, and the Quest for a New Utopia* (Audible Studios 2013).

Townsend D, Knoefel F and Goubran R, 'Privacy versus Autonomy: A Tradeoff Model for Smart Home Monitoring Technologies' (Annual International Conference of the IEEE Engineering in Medicine and Biology Society, Boston, Mass, USA, 30 August 2011).

Turow J, *The Aisles Have Eyes: How Retailers Track Your Shopping, Strip Your Privacy, and Define Your Power* (Yale University Press 2017).

van der Sloot B and Zuiderveen Borgesius F, 'The EU Data Protection Regulation: A New Global Standard for Information Privacy' <https://bartvandersloot.com/onewebmedia/SSRN-id3162987.pdf>.

van Dijck J, 'Datafication, Dataism and Dataveillance: Big Data between Scientific Paradigm and Ideology' (2014) 12 *Surveillance & Society* 197.

van Hoboken J, 'From Collection to Use in Privacy Regulation? A Forward-Looking Comparison of European and US Frameworks for Personal Data Processing' in van der Sloot B, Broeders D and Schrijvers E (eds), *Exploring the Boundaries of Big Data* (Amsterdam University Press 2016).

Vaughn A and others, 'Activity Detection and Analysis Using Smartphone Sensors' (2018 IEEE International Conference on Information Reuse and Integration (IRI), Salt Lake City, July 2018).

Volosovich S, 'Insurtech: Challenges and Development Perspectives' (2016) 3 *International Journal of Innovative Technologies in Economy* 39.

Waber B, *People Analytics: How Social Sensing Technology Will Transform Business and What It Tells Us about the Future of Work* (FT Press 2013).

Wachter S, Mittelstadt B and Floridi L, 'Why a Right to Explanation of Automated Decision-Making Does Not Exist in the General Data Protection Regulation' (2017) 7 *International Data Privacy Law* 76.

Wacks R, *Personal Information: Privacy and the Law* (Clarendon Press 1993).

Waldo J, Lin H and Millett LI, *Engaging Privacy and Information Technology in a Digital Age* (Information Privacy, National Academies Press 2007).

Warren SD and Brandeis LD, 'The Right to Privacy' (1890) 4 *Harvard Law Review* 193.

Weber S and Wong RY, 'The New World of Data: Four Provocations on The Internet of Things' (2017) 22 *First Monday* 1.

Westin AF, *Privacy and Freedom* (Atheneum 1967).

Whitson JR, 'Surveillance and Democracy in the Digital Enclosure' in Haggerty KD and Samatas M (eds), *Surveillance and Democracy* (Routledge, Taylor & Francis Group 2010).

Wilson C, Hargreaves T and Hauxwell-Baldwin R, 'Smart Homes and Their Users: A Systematic Analysis and Key Challenges' (2015) 19 *Personal and Ubiquitous Computing* 463.

—'Benefits and Risks of Smart Home Technologies' (2017) 103 *Energy Policy* 72.

Wilson JD, *Creating Strategic Value Through Financial Technology* (John Wiley & Sons 2017).

Witzleb N and Wagner J, 'When Is Personal Data about or Relating to an Individual? A Comparison of Australian, Canadian, and EU Data Protection and Privacy Laws' (2018) 4 *Canadian Journal of Comparative and Contemporary Law* 293.

World Economic Forum, *Personal Data: The Emergence of a New Asset Class* (World Economic Forum and Bain & Company 2011).

—*Big Data, Big Impact: New Possibilities for International Development* (2012).

Wu J and Coggeshall S, *Foundations of Predictive Analytics* (CRC Press 2012).

Yan TC, Schulte P and Lee DCK, 'InsurTech and FinTech: Banking and Insurance Enablement' in Lee DCK and Deng RH (eds), *Handbook of Blockchain, Digital Finance, and Inclusion* (Academic Press 2018), 249.

Yeung K, '"Hypernudge": Big Data as a Mode of Regulation by Design' (2017) 20 *Information, Communication & Society* 118.

—'Algorithmic Regulation: A Critical Interrogation' (2018) 12 *Regulation & Governance* 505.

—'Five Fears about Mass Predictive Personalization in an Age of Surveillance Capitalism' (2018) 8 *International Data Privacy Law* 258.

Zarsky TZ, 'Transparent Predictions' [2013] University of Illinois Law Review 1503.

—'The Privacy-Innovation Conundrum' (2015) 19 *Lewis & Clark Law Review* 168.

—'Incompatible: The GDPR in the Age of Big Data' (2017) 47 *Seton Hall Law Review* 1020.

—'Privacy and Manipulation in the Digital Age' (2019) 20 *Theoretical Inquiries in Law* 1.

Zeng E, Mare S and Roesner F, 'End User Security & Privacy Concerns with Smart Homes' (Symposium on Usable and Scable Privacy (SOUPS), Santa Clara, California, 2017).

Zhang J and others, 'WiFi-ID: Human Identification Using WiFi Signal' (2016 International Conference on Distributed Computing in Sensor Systems (DCOSS), 26–28 May 2016).

Zhang X and others, 'MoodExplorer: Towards Compound Emotion Detection via Smartphone Sensing' (2018) [ACM] 1 Proceedings of the ACM on Interactive, Mobile, Wearable and Ubiquitous Technologies 1.

Zittrain J, 'Privacy 2.0' [2008] The University of Chicago Legal Forum 65.

Zuboff S, 'Big Other: Surveillance Capitalism and the Prospects of an Information Civilization' (2015) 30 *Journal of Information Technology* 75.

—*The Age of Surveillance Capitalism: The Fight for a Human Future at the New Frontier of Power* (1st edn, Public Affairs 2019).

Zuiderveen Borgesius FJ, 'Singling Out People without Knowing Their Names: Behavioural Targeting, Pseudonymous Data, and the New Data Protection Regulation' (2016) 32 *Computer Law & Security Review: The International Journal of Technology Law and Practice* 256.

Index

Cambridge Intellectual Property and Information Law

Titles in the Series (formerly known as Cambridge Studies in Intellectual Property Rights)

Printed in the United States
by Baker & Taylor Publisher Services